Decision Trees for Analytics
Using SAS® Enterprise Miner™

Barry de Ville and Padraic Neville

support.sas.com/bookstore

The correct bibliographic citation for this manual is as follows: de Ville, Barry, and Padraic Neville. 2013. *Decision Trees for Analytics Using SAS® Enterprise Miner*. Cary, NC: SAS Institute Inc.

Decision Trees for Analytics Using SAS® Enterprise Miner

ISBN 978-1-61290-252-4 (electronic book)
ISBN 978-1-61290-315-6

SAS Institute Inc., SAS Campus Drive, Cary, North Carolina 27513-2414

1st printing, June 2013

Contents

Preface

This updated book on decision trees combines the talents and knowledge of two of the most experienced decision tree practitioners in the field today. Barry de Ville was a pioneer in the implementation of the CHAID and XAID approaches advocated by Kass (1975, 1980) and Hawkins and Kass (1982). He led the development of the first commercial implementation of an integrated CHAID/XAID approach to decision trees (deVille, 1990). Padraic Neville began developing decision tree algorithms in 1983 and served on the development team for the commercial implementation of the approach described in *Classification and Regression Trees* (Breiman et al. 1984). Padraic has served as the primary developer of the SAS decision tree software since the inauguration of the procedure in the first SAS Enterprise Miner release in 1999. He has seamlessly and artfully blended the foundational decision tree traditions together in the development of an integrated package that offers the best of all traditions in one place. This creates a unique ability to mix and match approaches, to make the appropriate adjustments, and to create an optimal decision tree.

In addition to providing an exhaustive treatment of the end-to-end process of decision tree construction and the respective considerations and algorithms—as provided in the first edition—this edition adds up-to-date treatments of boosting and forest approaches, rule induction, and a dedicated section on the most recent findings related to bias reduction in variable selection.

The coverage includes discussions of key issues in decision tree practice: how to protect against overfitting by proactively adjusting p-values (from the CHAID/XAID tradition) or by retrospectively pruning with validation (or cross-validation) in the Breiman et al. tradition. In the same fashion, multiple methods of dealing with missing values are described (treat missing as valid, assign it to the closest branch, use a surrogate, or distribute the values across branches). The various aspects of these approaches are all covered here, in one place. The overall framework that enables the user to incorporate prior probabilities or costs in the split search, pruning, and tree formation is also discussed.

New additions to the SAS Enterprise Miner decision tree capabilities, such as support for multiple targets and switching targets mid-tree, are also provided. Additional sections on recent developments in rule induction are included.

As with the first edition, this current edition remains the most comprehensive treatment of decision tree theory, use, and applications available in one easy-to-access place.

About This Book

Purpose

We wrote this book to familiarize people with the use and operation of decision trees in a range of data analysis and data consumption tasks and situations.

Is This Book for You?

If you are interested in understanding the powerful and popular approach of using decision trees for the analysis and presentation of data then you will find this book useful.

Prerequisites

There are minimal prerequisites to use this book productively. We offer many illustrations and explanations of decision tree operation ranging from the very simple to the reasonably advanced. Very little formal mathematical notation is used; instead, general-purpose, high-school level English is used in all our descriptions and explanations.

Organization of This Book

This book begins with a general introduction to decision trees and includes background history and the evolution of the conceptual underpinnings and major tools of decision tree construction. Intermediate sections describe data preparation considerations, data assembly within the decision tree product environment, as well as the production of useful and understandable results. Later sections provide some topics that may be considered more advanced or esoteric: displays that feature multiple targets, the relationship of decision trees to machine learning, multi-tree "ensembles" and variations of decision trees in unique predictive and specialized application areas are discussed.

Software Used to Develop This Book's Content

We used the Decision Tree modeling node of SAS Enterprise Miner to develop all the examples used in this book. No new features beyond the features available in the 6.1 version of SAS Enterprise Miner are explicitly used; however, some sections describe the latest developments as of the 12.1 release, dated 2013 (the year of publication of this edition).

Author Pages

You can access the two author pages for this book at

- http://support.sas.com/publishing/authors/deville.html
- http://support.sas.com/publishing/authors/neville.html.

The author pages include all of the SAS Press books that have been written by these authors. For features that relate to this book, look for the cover image of this book. The links below the book cover will take you to a free chapter, example code and data, reviews, updates, and more.

Example Code and Data

You can access the example code and data for this book by linking to its author pages at http://support.sas.com/publishing/authors.

Select the name of the author. Then look for the cover image of this book and select the "Example Code and Data" link to display the SAS programs that are included in this book.

For an alphabetical listing of all books for which example code and data is available, see http://support.sas.com/bookcode. Select a title to display the book's example code.

If you are unable to access the code through the web site, send e-mail to saspress@sas.com.

Additional Resources

SAS offers you a rich variety of resources to help build your SAS skills and explore and apply the full power of SAS software. Whether you are in a professional or academic setting, we have learning products that can help you maximize your investment in SAS.

Bookstore	http://support.sas.com/bookstore/
Training	http://support.sas.com/training/
Certification	http://support.sas.com/certify/
SAS Global Academic Program	http://support.sas.com/learn/ap/
SAS OnDemand	http://support.sas.com/learn/ondemand/
Knowledge Base	http://support.sas.com/resources/
Support	http://support.sas.com/techsup/
Training and Bookstore	http://support.sas.com/learn/
Community	http://support.sas.com/community/

Keep in Touch

We look forward to hearing from you. We invite questions, comments, and concerns. If you want to contact us about a specific book, please include the book title in your correspondence.

To Contact the Authors through SAS Press

By e-mail: saspress@sas.com

Via the Web: http://support.sas.com/author_feedback

SAS Books

For a complete list of books available through SAS visit http://support.sas.com/bookstore.

Phone: 1-800-727-3228

Fax: 1-919-677-8166

E-mail: sasbook@sas.com

SAS Book Report

Receive up-to-date information about all new SAS publications via e-mail by subscribing to the SAS Book Report monthly eNewsletter. Visit http://support.sas.com/sbr.

About These Authors

Barry de Ville is a Solutions Architect at SAS. His work with decision trees has been featured during several SAS users' conferences and has led to the award of a U.S. patent on "bottom-up" decision trees. Previously, Barry led the development of the KnowledgeSEEKER decision tree package. He has given workshops and tutorials on decision trees at such organizations as Statistics Canada, the American Marketing Association, the IEEE, and the Direct Marketing Association.

Padraic Neville is a Principal Research Statistician Developer at SAS. He developed the decision tree and boosting procedures in SAS Enterprise Miner and the high-performance procedure HPFOREST. In 1984 Padraic produced the first commercial release of the Classification and Regression Trees software by Breiman, Friedman, Olshen, and Stone. He since has taught decision trees at the Joint Statistical Meetings. His current research pursues better insight and prediction with multiple trees.

Learn more about these authors by visiting their author pages where you can download free chapters, access example code and data, read the latest reviews, get updates, and more:

- http://support.sas.com/deville
- http://support.sas.com/neville

About These Authors

Acknowledgments

The authors would like to acknowledge and thank the many users, developers and advocates of decision tree methods who have consistently shared support, encouragement and enthusiasm throughout the years. We share a common legacy where all viewpoints are welcome guidelines to a continuously evolving future.

A number of close colleagues have been helpful and supportive in the course of the development of decision trees at SAS (now in its second decade), most especially Warren Sarle and Pei-Yi Tan.

We are most thankful for the production team that makes any publication possible and reserve our deepest gratitude for the enduring support, advice and assistance from our publishing colleagues – notably Julie Platt and John West.

Chapter 1: Decision Trees—What Are They?

Introduction

Decision trees are a simple, but powerful form of multiple variable analysis. They provide unique capabilities to supplement, complement, and substitute for:

- traditional statistical forms of analysis (such as multiple linear regression)
- a variety of data mining tools and techniques (such as neural networks)
- recently developed multidimensional forms of reporting and analysis found in the field of business intelligence

Decision trees are produced by algorithms that identify various ways of splitting a data set into branch-like segments. These segments form an inverted decision tree that originates with a root node at the top of the tree. The object of analysis is reflected in this root node as a simple, one-dimensional display in the decision tree interface. The name of the field of data that is the object of analysis is usually displayed, along with the spread or distribution of the values that are contained in that field. A sample decision tree is illustrated in Figure 1.1, which shows that the decision tree can reflect both a continuous and categorical object of analysis. The display of this node reflects all the data set records, fields, and field values that are found in the object of analysis. The discovery of the decision rule to form the branches or segments underneath the root node is based on a method that extracts the relationship between the object of analysis (that serves as the target field in the data) and one or more fields that serve as input fields to create the branches or segments. The values in the input field are used to estimate the likely value in the target field. The target field is also called an outcome, response, or dependent field or variable.

The general form of this modeling approach is illustrated in Figure 1.1. Once the relationship is extracted, then one or more decision rules that describe the relationships between inputs and targets can be derived. Rules can be selected and used to display the decision tree, which provides a means to visually examine and describe the tree-like network of relationships that characterize the input and target values. Decision rules can predict the values of new or unseen observations that contain values for the inputs, but that might not contain values for the targets.

Figure 1.1: Illustration of the Decision Tree

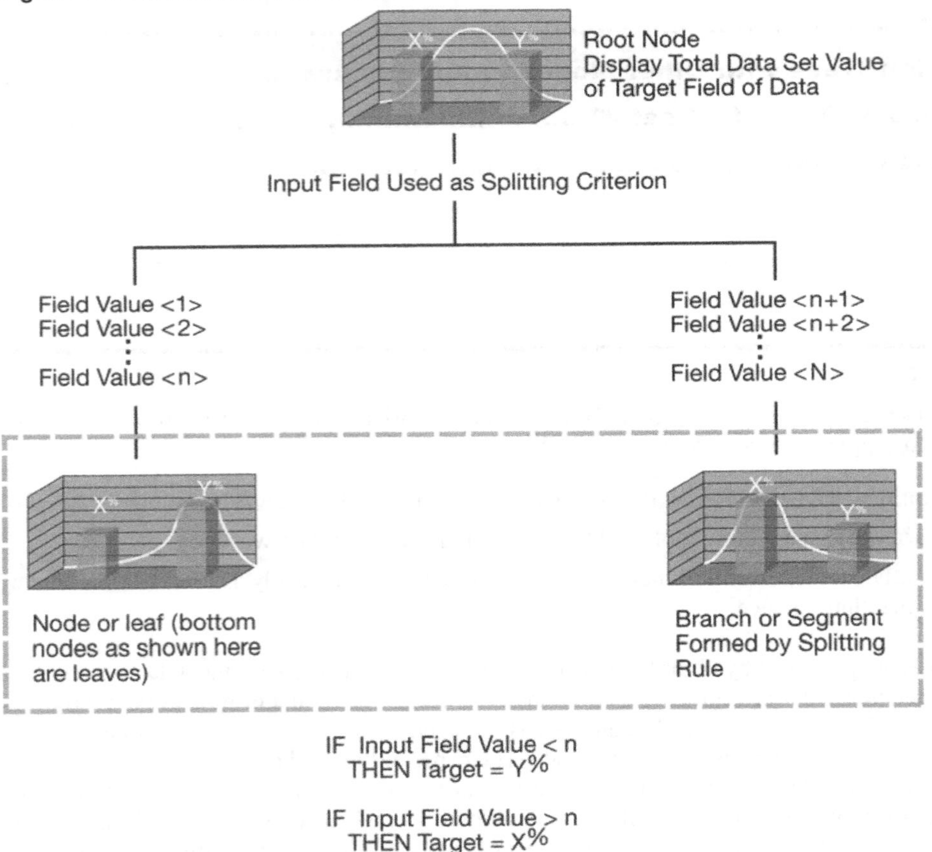

Each rule assigns a record or observation from the data set to a node in a branch or segment based on the value of one of the fields or columns in the data set.[1] Fields or columns that are used to create the rule are called *inputs*. Splitting rules are applied one after another, resulting in a hierarchy of branches within branches that produces the characteristic inverted decision tree form. The nested hierarchy of branches is called a *decision tree*, and each segment or branch is called a *node*. A node with all its descendent segments forms an additional segment or a branch of that node. The bottom nodes of the decision tree are called *leaves* (or *terminal nodes*). For each leaf, the decision rule provides a unique path for data to enter the class that is defined as the leaf. All nodes, including the bottom leaf nodes, have mutually exclusive assignment rules. As a result, records or observations from the parent data set can be found in one node only. Once the decision rules have been determined, it is possible to use the rules to predict new node values based on new or unseen data. In predictive modeling, the decision rule yields the predicted value.

Figure 1.2: Illustration of Decision Tree Nomenclature

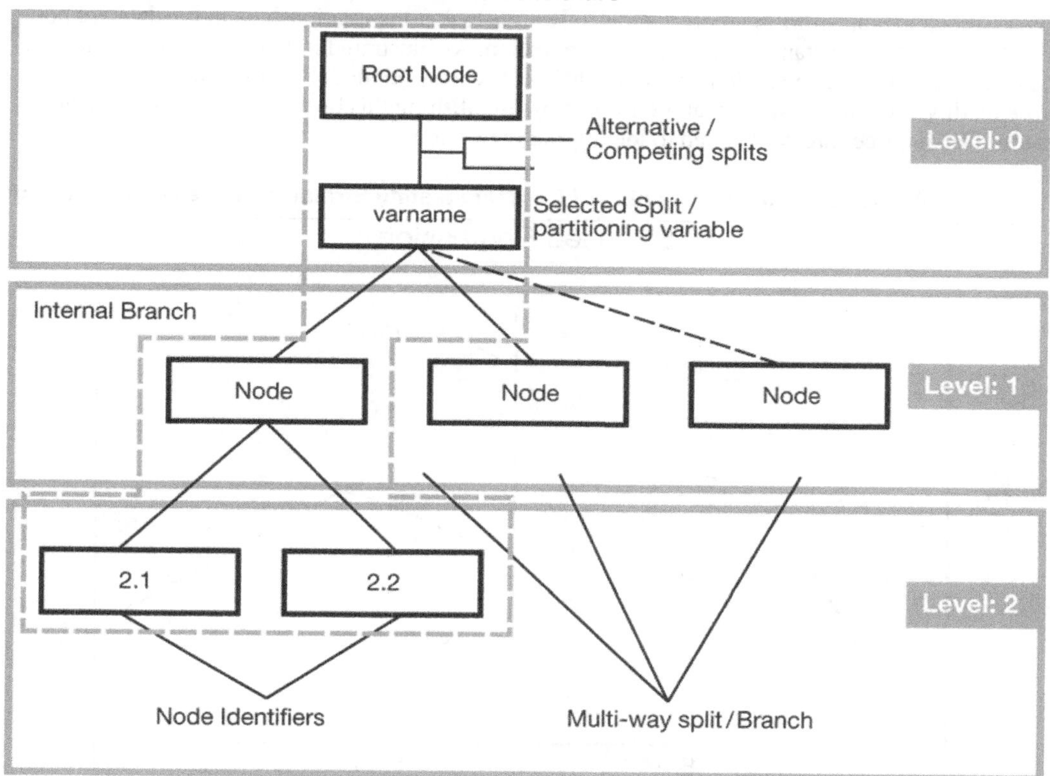

Although decision trees have been in development and use for over 50 years (one of the earliest uses of decision trees was in the study of television broadcasting by Belson in 1956), many new forms of decision trees are evolving that promise to provide exciting new capabilities in the areas of data mining and machine learning in the years to come. For example, one new form of the decision tree involves the creation of *random forests*. Random forests are multi-tree committees that use randomly drawn samples of data and inputs and reweighting techniques to develop multiple trees that, when combined, provide for stronger prediction and better diagnostics on the structure of the decision tree.

Besides modeling, decision trees can be used to explore and clarify data for dimensional cubes that are found in business analytics and business intelligence.

Using Decision Trees with Other Modeling Approaches

Decision trees play well with other modeling approaches, such as regression, and can be used to select inputs or to create dummy variables representing interaction effects for regression equations. For example, Neville (1998) explains how to use decision trees to create stratified regression models by selecting different slices of the data population for in-depth regression modeling.

The essential idea in stratified regression is to recognize that the relationships in the data are not readily fitted for a constant, linear regression equation. As illustrated in Figure 1.3, a boundary in the data could suggest a partitioning so that different regression models of different forms can be more readily fitted in the strata that are formed by establishing this boundary. As Neville (1998) states, decision trees are well suited to identifying regression strata.

Figure 1.3: Illustration of the Partitioning of Data Suggesting Stratified Regression Modeling

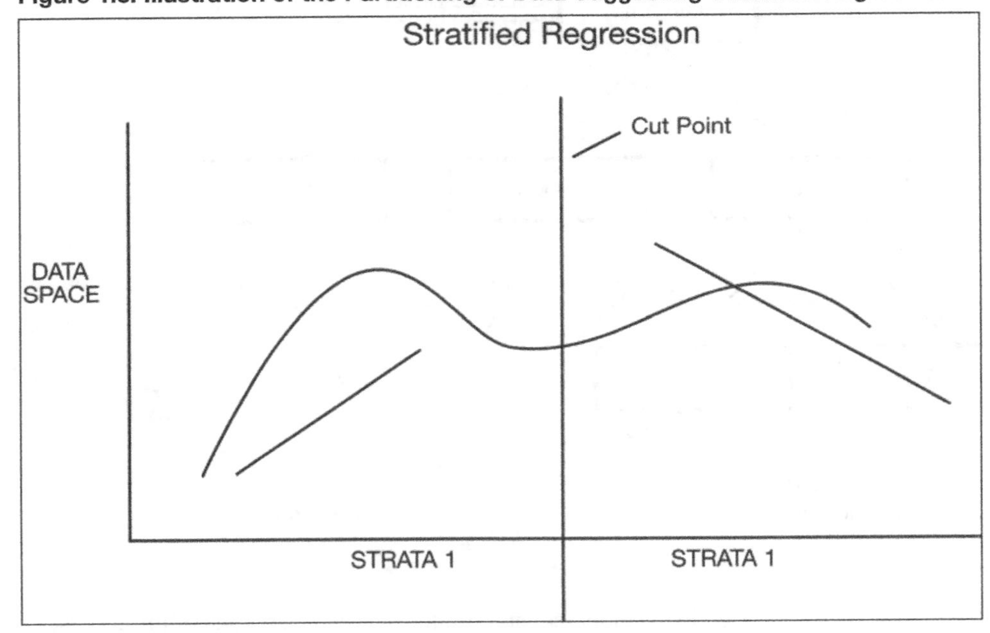

Decision trees are also useful for collapsing a set of categorical values into ranges that are aligned with the values of a selected target variable or value. This is sometimes called *optimal collapsing of values*. A typical way of collapsing categorical values together would be to join adjacent categories together. In this way 10 separate categories can be reduced to 5. In some cases, as illustrated in Figure 1.4, this results in a significant reduction in information. Here, categories 1 and 2 are associated with extremely low and extremely high levels of the target value. In this example, the collapsed categories 3 and 4, 5 and 6, 7 and 8, and 9 and 10 work better in this type of deterministic collapsing framework; however, the anomalous outcome produced by collapsing categories 1 and 2 together should serve as a strong caution against adopting any such scheme on a regular basis.

Decision trees produce superior results. The dotted lines show how collapsing the categories with respect to the levels of the target yields different and better results. If we impose a monotonic restriction on the collapsing of categories—as we do when we request tree growth on the basis of ordinal predictors—then we see that category 1 becomes a group of its own. Categories 2, 3, and 4 join together and point to a relatively high level in the target. Categories 5, 6, and 7 join together to predict the lowest level of the target. And categories 8, 9, and 10 form the final group.

If a completely unordered grouping of the categorical codes is requested—as would be the case if the input was defined as "nominal"—then the three bins as shown at the bottom of Figure 1.4 might be produced. Here, the categories 1, 5, 6, 7, 9, and 10 group together as associated with the highest level of the target. The medium target levels produce a grouping of categories 3, 4, and 8. The lone high target level that is associated with category 2 falls out as a category of its own.

Figure 1.4: Illustration of Forming Nodes by Binning Input-Target Relationships

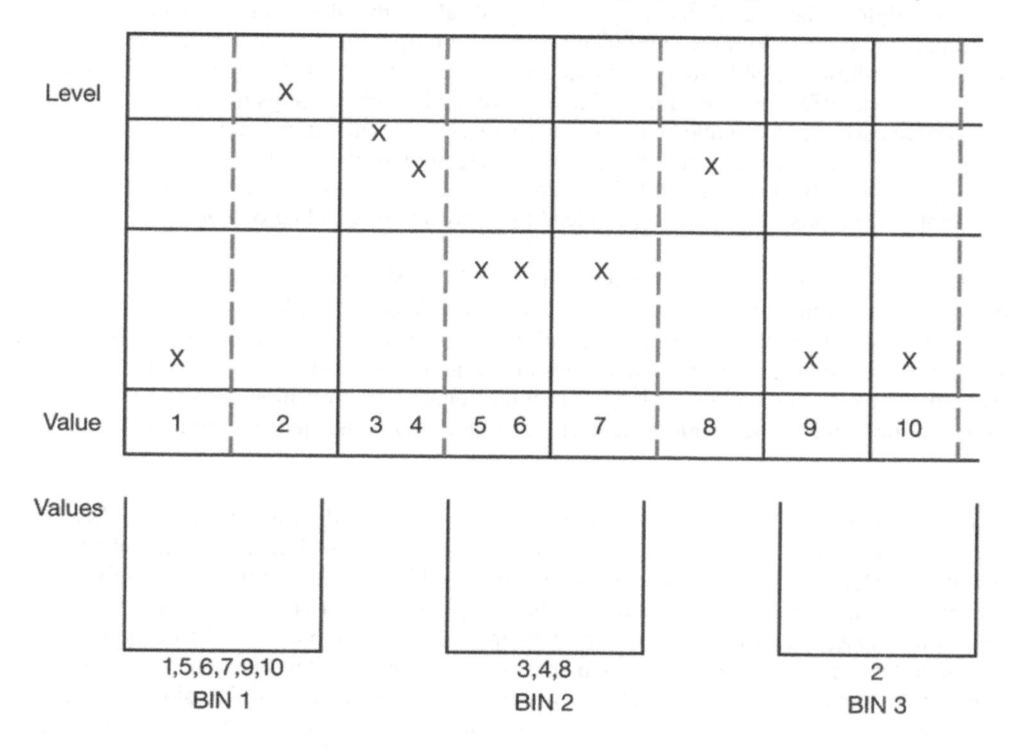

Because a decision tree enables you to combine categories that have similar values with respect to the level of some target value, there is less information loss in collapsing categories together. This leads to improved prediction and classification results. As shown in the figure, it is possible to intuitively appreciate that these collapsed categories can be used as branches in a tree. So, knowing the branch—for example, branch 3 (labeled BIN 3), we are better able to guess or predict the level of the target. In the case of branch 2, we can see that the target level lies in the mid-range, whereas in the last branch—here collapsed categories 1, 5, 6, 7, 9, 10—the target is relatively low.

Why Are Decision Trees So Useful?

Decision trees are a form of multiple variable (or multiple effect) analyses. All forms of multiple variable analyses enable us to predict, explain, describe, or classify an outcome (or target). An example of a multiple variable analysis is a probability of sale or the likelihood to respond to a marketing campaign as a result of the combined effects of multiple input variables, factors, or dimensions. This multiple variable analysis capability of decision trees enables you to go beyond simple one-cause, one-effect relationships and to discover and describe things in the context of multiple influences. Multiple variable analysis is particularly important in current problem-solving because almost all critical outcomes that determine success are based on multiple factors. Further, it is becoming increasingly clear that while it is easy to set up one-cause, one-effect relationships in the form of tables or graphs, this approach can lead to costly and misleading outcomes.

According to research in cognitive psychology (Miller 1956; Kahneman, Slovic, and Tversky 1982) the ability to conceptually grasp and manipulate multiple chunks of knowledge is limited by the physical and cognitive processing limitations of the short-term memory portion of the brain. This places a premium on the utilization of dimensional manipulation and presentation techniques that are capable of preserving and reflecting high-dimensionality relationships in a readily comprehensible form so that the relationships can be more easily consumed and applied by humans.

There are many multiple variable techniques available. The appeal of decision trees lies in their relative power, ease of use, robustness with a variety of data and levels of measurement, and ease of interpretability. Decision trees are developed and presented incrementally; thus, the combined set of multiple influences (which are necessary to fully explain the relationship of interest) is a collection of one-cause, one-effect relationships presented in the recursive form of a decision tree. This means that decision trees deal with human short-term memory limitations quite effectively and are easier to understand than more complex, multiple variable techniques. Decision trees turn raw data into an increased knowledge and awareness of business, engineering, and scientific issues, and they enable you to deploy that knowledge in a simple but powerful set of human-readable rules.

Decision trees attempt to find a strong relationship between input values and target values in a group of observations that form a data set. When a set of input values is identified as having a strong relationship to a target value, all of these values are grouped in a bin that becomes a branch on the decision tree. These groupings are determined by the observed form of the relationship between the bin values and the target. For example, suppose that the target average value differs sharply in the three bins that are formed by the input. As shown in Figure 1.4, binning involves taking each input, determining how the values in the input are related to the target, and, based on the input-target relationship, depositing inputs with similar values into bins that are formed by the relationship.

To visualize this process using the data in Figure 1.4, you see that BIN 1 contains values 1, 5, 6, 7, 9, and 10; BIN 2 contains values 3, 4, and 8; and BIN 3 contains value 2. The sort-selection mechanism can combine values in bins whether or not they are adjacent to one another (e.g., 3, 4, and 8 are in BIN 2, whereas 7 is in BIN 1). When only adjacent values are allowed to combine to form the branches of a decision tree, the underlying form of measurement is assumed to monotonically increase as the numeric code of the input increases. When non-adjacent values are allowed to combine, the underlying form of measurement is non-monotonic. A wide variety of different forms of measurement, including linear, nonlinear, and cyclic, can be modeled using decision trees.

A strong input-target relationship is formed when knowledge of the value of an input improves the ability to predict the value of the target. A strong relationship helps you understand the characteristics of the target. It is normal for this type of relationship to be useful in predicting the values of targets. For example, in most animal populations, knowing the height or weight of the individual improves the ability to predict the gender. In the following display, there are 28 observations in the data set. There are 20 males and 8 females.

Table 1.1: Age, Height, and Gender Relationships

bodytype	Gender	BMI	Weight	Height	Ht_Centimeters
slim	Female	162.389	179	4'10	147.32
slim	Female	161.275	160	5' 4	162.56
average	Male	181.630	191	5' 8	172.72
slim	Male	143.011	132	5'1	154.94
average	Female	173.542	167	5'1	180.34
slim	Female	141.977	128	5'2	157.48
slim	Female	153.695	150	5'2	157.48
slim	Male	153.695	150	5'2	157.48
heavy	Female	184.006	215	5'2	157.48
slim	Female	119.339	89	5'3	160.02
slim	Female	163.473	167	5'3	160.02
average	Male	171.058	180	5'4	162.56
average	Male	182.996	206	5'4	162.56
heavy	Male	198.643	239	5'5	165.10
average	Male	164.286	161	5'6	167.64
average	Male	177.528	188	5'6	167.64
heavy	Male	218.197	284	5'6	167.64
slim	Female	141.107	117	5'7	170.18
average	Male	166.551	163	5'7	170.18
average	Male	181.700	194	5'7	170.18
heavy	Male	184.949	201	5'7	170.18
heavy	Male	209.454	254	5'8	172.72
heavy	Male	187.689	201	5'9	175.26
heavy	Male	190.009	206	5'9	175.26
heavy	Male	194.567	216	5'9	175.26
heavy	Male	194.096	206	6'	182.88
heavy	Male	201.971	220	6'1	185.42
heavy	Female	184.956	182	6'2	187.96

In this display, the overall average height is 5'6 and the overall average weight is 183. Among males, the average height is 5'7, while among females, the average height is 5'3 (males weigh 200 on average, versus 155 for females).

Knowing the gender puts us in a better position to predict the height and weight of the individuals, and knowing the relationship between gender and height and weight puts us in a better position to understand the characteristics of the target. Based on the relationship between height and weight and gender, you can infer that females are both smaller and lighter than males. As a result, you can see how this sort of knowledge that is based on gender can be used to determine the height and weight of unseen humans.

From the display, you can construct a branch with three leaves to illustrate how decision trees are formed by grouping input values based on their relationship to the target.

Figure 1.5: Illustration of Decision Tree Partitioning of Physical Measurements

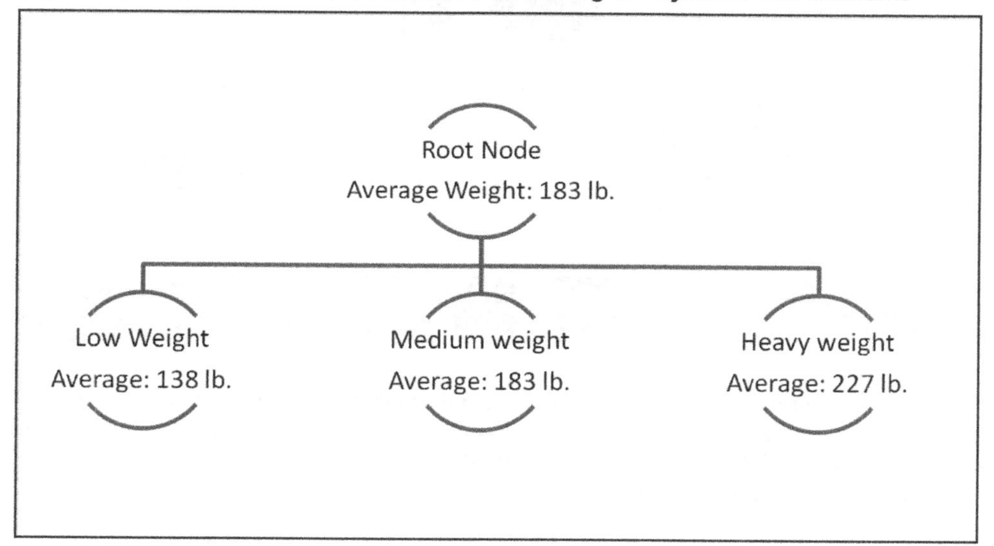

Level of Measurement

The example shown here illustrates an important characteristic of decision trees: both quantitative and qualitative data can be accommodated in decision tree construction. Quantitative data, like height and weight, refers to quantities that can be manipulated with arithmetic operations such as addition, subtraction, and multiplication. Qualitative data, such as gender, cannot be used in arithmetic operations, but can be presented in tables or decision trees. In the previous example, the target field is weight and is presented as an average. Height, BMIndex, or BodyType could have been used as inputs to form the decision tree.

Some data, such as shoe size, behaves like both qualitative and quantitative data. For example, you might not be able to do meaningful arithmetic with shoe size, even though the sequence of numbers in shoe sizes is in an observable order. For example, with shoe size, size 10 is larger than size 9, but it is not twice as large as size 5.

Figure 1.6 displays a decision tree developed with a categorical target variable. This figure shows the general, tree-like characteristics of a decision tree and illustrates how decision trees display multiple relationships—one branch at a time. In subsequent figures, decision trees are shown with continuous or numeric fields as targets. This shows how decision trees are easily developed using targets and inputs that are both qualitative (categorical data) and quantitative (continuous, numeric data).

Figure 1.6: Illustration of a Decision Tree with a Categorical Target

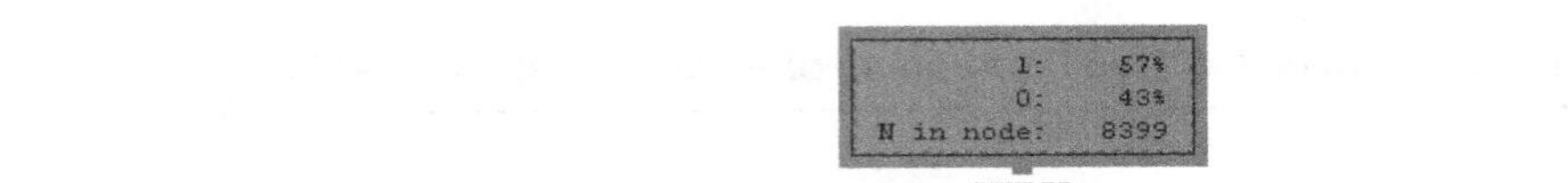

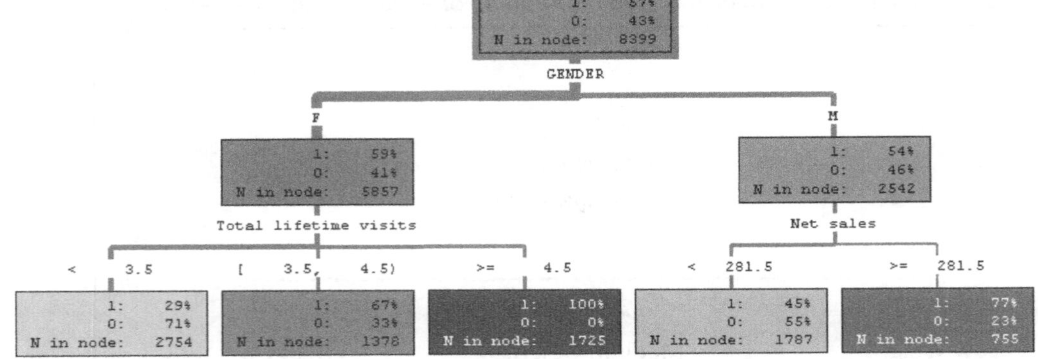

The decision tree in Figure 1.6 displays the results of a mail-in customer survey conducted by HomeStuff, a national home goods retailer. In the survey, customers had the option to enter a cash drawing. Those who entered the drawing were classified as a HomeStuff *best* customer. Best customers are coded with **1** in the decision tree.

The top-level node of the decision tree shows that, of the 8399 respondents to the survey, 57% were classified as best customers, while 43% were classified as *other* (coded with **0**).

Figure 1.6 shows the general characteristics of a decision tree, such as partitioning the results of a 1–0 (categorical) target across various input fields in the customer survey data set. Under the top-level node, the field **GENDER** further characterizes the best – other (1–0) response. Females (coded with **F**) are more likely to be best customers than males (coded with **M**). Fifty-nine percent of females are best customers versus 54% of males. A wide variety of splitting techniques have been developed over time to gauge whether this difference is statistically significant and whether the results are accurate and reproducible. In Figure 1.6, the difference between males and females is statistically significant. Whether a difference of 5% is significant from a business point of view is a question that is best answered by the business analyst.

The splitting techniques that are used to split the 1–0 responses in the data set are used to identify alternative inputs (for example, income or purchase history) for gender. These techniques are based on numerical and statistical techniques that show an improvement over a simple, uninformed guess at the value of a target (in this example, best–other), as well as the reproducibility of this improvement with a new set of data.

Knowing the gender enables us to guess that females are 5% more likely to be a best customer than males. You could set up a separate, independent *hold-out* or *validation* data set, and (having determined that the gender effect is useful or interesting) you might see whether the strength and direction of the effect is reflected in the hold-out or validation data set. The separate, independent data set will show the results if the decision tree is applied to a new data set, which indicates the generality of the results. Another way to assess the generality of the results is to look at data distributions that have been studied and developed by statisticians who know the properties of the data and who have developed guidelines based on the properties of the data and data distributions. The results could be compared to these data distributions and, based on the comparisons, you could determine the strength and reproducibility of the results. These approaches are discussed at greater length in Chapter 3, "The Mechanics of Decision Tree Construction."

Under the female node in the decision tree in Figure 1.6, female customers can be further categorized into best–other categories based on the total lifetime visits that they have made to HomeStuff stores. Those who have made fewer than 3.5 visits are less likely to be best customers compared to those who have made more than 4.5 visits: 29% versus 100%. (In the survey, a shopping visit of less than 20 minutes was characterized as a half visit.)

On the right side of the figure, the decision tree is asymmetric; a new field—**Net sales**—has entered the analysis. This suggests that **Net sales** is a stronger or more relevant predictor of customer status than **Total lifetime visits**, which was used to analyze females. It was this kind of asymmetry that spurred the initial development of decision trees in the statistical community: these kinds of results demonstrate the importance of the combined (or interactive) effect of two indicators in displaying the drivers of an outcome. In the case of males, when net sales exceed $281.50, then the likelihood of being a best customer increases from 45% to 77%.

As shown in the asymmetry of the decision tree, female behavior and male behavior have different nuances. To explain or predict female behavior, you have to look at the interaction of gender (in this case, female) with **Total lifetime visits**. For males, **Net sales** is an important characteristic to look at.

In Figure 1.6, of all the k-way or n-way branches that could have been formed in this decision tree, the 2-way branch is identified as best. This indicates that a 2-way branch produces the strongest effect. The strength of the effect is measured through a criterion that is based on strength of separation, statistical significance, or reproducibility, with respect to a validation process. These measures, as applied to the determination of branch formation and splitting criterion identification, are discussed further in Chapter 3.

Decision trees can accommodate categorical (gender), ordinal (number of visits), and continuous (net sales) types of fields as inputs or classifiers for the purpose of forming the decision tree. Input classifiers can be created by binning quantitative data types (ordinal and continuous) into categories that might be used in the creation of branches—or splits—in the decision tree. The bins that form total lifetime visits have been placed into three branches:

- < 3.5 … less than 3.5
- [3.5 – 4.5) … between 3.5 to strictly less than 4.5
- >= 4.5 … greater than or equal to 4.5

Various nomenclatures are used to indicate which values fall in a given range. Meyers (1990) proposes the following alternative:

- < 3.5 … less than 3.5
- [3.5 – 4.5[… between 3.5 to strictly less than 4.5
- >= 4.5 … greater than or equal to 4.5

The key difference between these alternatives and the convention used in the SAS decision tree is in the second range of values, where the bracket designator ([) is used to indicate the interval that includes the lower number and includes up to any number that is strictly less than the upper number in the range.

A variety of techniques exist to cast bins into branches: 2-way (binary branches), n-way (where **n** equals the number of bins or categories), or k-way (where **k** represents an attempt to create an optimal number of branches and is some number greater than or equal to 2 and less than or equal to n).

Figure 1.7: Illustration of a Decision Tree—Continuous (Numeric) Target

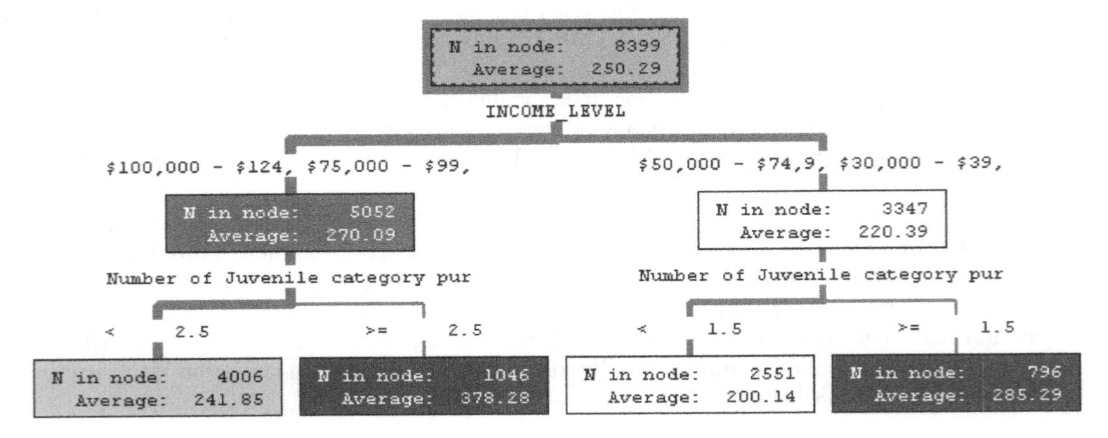

Figure 1.7 shows a decision tree that is created with a continuous response variable as the target. In this case, the target field is **Net sales**. This is the same field that was used as a classifier (for males) in the categorical response decision tree shown in Figure 1.6.

Overall, as shown in Figure 1.7, the average net sale amount is approximately $250. Figure 1.7 shows how this amount can be characterized by performing successive splits of net sales according to the income level of the survey responders and, within their income level, according to the field **Number of Juvenile category purchases**. In addition to characterizing net sales spending groups, this decision tree can be used as a predictive tool. For example, in Figure 1.7, high income, high juvenile category purchases typically outspend the average purchaser by an average of $378, versus the norm of $250. If someone were to ask what a relatively low income purchaser who buys a relatively low number of juvenile category items would spend, then the best guess would be about $200. This result is based on the decision rule, taken from the decision tree, as follows:

```
IF Number of Juvenile category purchases       <     1.5
      AND INCOME_LEVEL     $50,000 - $74,9,
                           $40,000 - $49,9,
                           $30,000 - $39,9,
                               UNDER $30,000
                           THEN Average Net Sales = $200.14
```

Decision trees can contain both categorical and numeric (continuous) information in the nodes of the tree. Similarly, the characteristics that define the branches of the decision tree can be both categorical or numeric (in the latter case, the numeric values are collapsed into bins—sometimes called buckets or collapsed groupings of categories—to enable them to form the branches of the decision tree).

Figure 1.8 shows how the Fisher-Anderson iris data can yield three different types of branches when classifying the target SETOSA versus OTHER (Fisher 1936). In this case, there are 2-, 3-, and 5-leaf branches. There are 50 SETOSA records in the data set. With the binary partition, these records are classified perfectly by the rule **petal width <= 6 mm**. The 3-way and 5-way branch partitions are not as effective as the 2-way partition and are shown only for illustration. More examples are provided in Chapter 2, "Descriptive, Predictive, and Explanatory Analyses," including examples that show how 3-way and n-way partitions are better than 2-way partitions.

Figure 1.8: Illustration of Fisher-Anderson Iris Data and Decision Tree Options

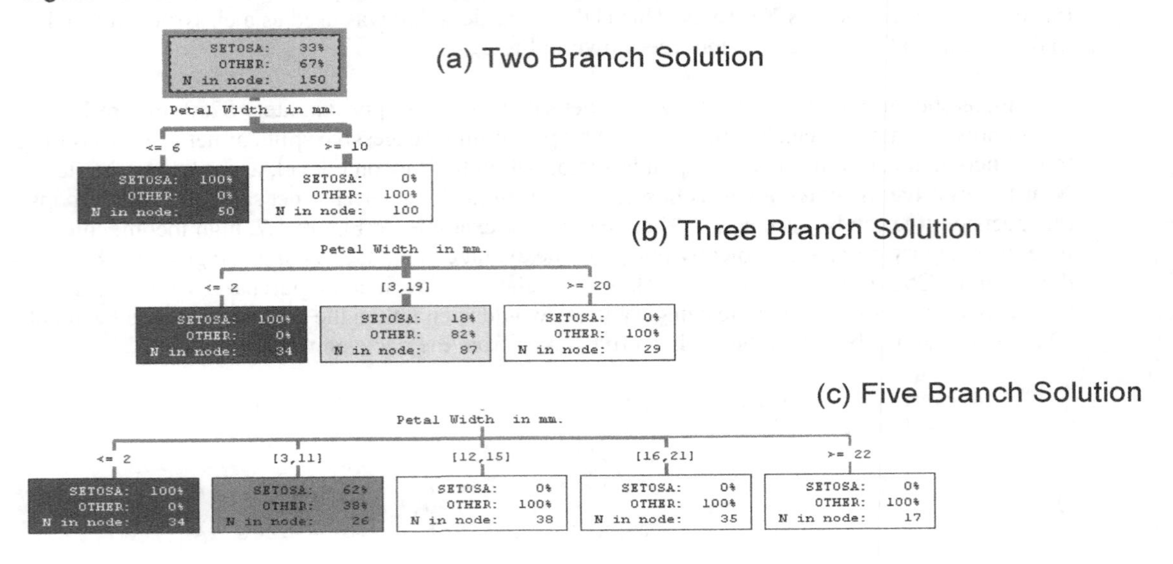

---

[1] The SAS Enterprise Miner decision tree contains a variety of algorithms to handle missing values, including a unique algorithm to assign partial records to different segments when the value in the field that is being used to determine the segment is missing.

Chapter 2: Descriptive, Predictive, and Explanatory Analyses

Introduction

In data analysis, it is common to work with data with descriptive, predictive, or explanatory outcomes in mind. A descriptive analysis could simply display a relationship in data or it could display the relationship as a graphic, such as a bar chart. The goal is to describe the data or a relationship among various data elements in the data set. This is common and is normally the baseline point of departure in working with data to develop insight. For example, you could describe the weather by indicating the temperature, relative humidity, or atmospheric pressure.

Predictive use of data is a little different from descriptive use of data. In the predictive setting, it is normal to describe a relationship among data elements; furthermore, you can assert that this relationship will hold over time and will be the same with new data, meaning that the relationship will be roughly reproduced in a novel situation. In the weather example, you can predict a weather effect based on the current rate of movement of a weather pattern, the differential pressure between competing weather systems, and air path measurements such as land mass, temperature, and humidity.

The explanatory use of data describes a relationship and attempts to show, by reference to the data, the effect and interpretation of the relationship. In the weather example, you could say that the effect of temperature on air mass humidity is rain or snow, depending on the degrees of temperature and the percent of humidity in the air (and other factors, such as atmospheric pressure and air particle concentration).

Typically, you must step up the rigor of the data work and task organization as you move from descriptive use to explanatory use. In a descriptive setting, the baseline goal is likely to be to present the facts in a clear and unambiguous fashion. In a predictive setting, the baseline goal is likely to be to produce a reliable and reproducible predicted outcome (which is usually confirmed by reference to validation or test data drawn from a novel, but related, set of circumstances as the host data used to train the predictive model). In a predictive setting, it is important to show the numerical relationship between predictive rules or equations and the target value. As a result, you can say that an increase in, for example, 10 units of a given predictor is likely to cause an increase in 2 units of the target or outcome of the prediction.

The explanatory use of data is more difficult to implement than either the descriptive or predictive use. Here, it is necessary to show how and to what degree a given relationship that is reflected in the data occurs. Usually, this demonstration is through reference to some explicit or implicit explanatory concept. For example, you can say that there is a direct relationship between air pressure and buoyancy of an air mass (or, for that matter, you can assert that there is a direct relationship between air pressure and the boiling temperature of water). Here, in the explanatory setting, you must show, through some kind of experiment, that the supposed relationship holds across various points of measurement, in different circumstances, and in different points in time. For example, if you describe the effect of air pressure on the boiling temperature of water, you might predict the boiling point at a given atmospheric pressure and then confirm the prediction through a measurement in an experimental setting. The most effective explanations demonstrate that the presumed relationship is primary, in that it is not an artifact of some preexisting relationship, nor is it mimicking the effects of an overarching or intervening relationship that is not expressed in the explanatory concept.

The Importance of Showing Context

Decision trees are constructed through successive recursive branches, where a branch is contained within the parent branch and is usually accompanied by peers that are formed at the same level of the decision tree. Because of this, a defining characteristic of a decision tree is that it clearly and graphically displays the interrelationships among the multiple factors that form the decision tree model, as viewed from branch to branch and between branches at any level of the decision tree. Decision trees display contextual effects—hot spots and soft spots in relationships that characterize data. These hot spots and soft spots reveal the frequently hidden and sometimes counterintuitive complexities in a relationship that unlock the decision-making potential of the data. For example, explore symmetry in branches that are peers at a given level of the decision tree: are sub-branches of a male gender split formed by the same inputs as sub-branches of a female gender split? In other words, are these relationships symmetrical? Is the direction of the relationship the same? Or, is there a reversal of the relationship—an interaction—that depends on the parent split?

You intuitively know the importance of multiple, contextual effects, but you often find it difficult to understand the context because of the inherent difficulty of capturing and describing the richly woven complexity of multiple, interrelated factors. It is tempting to resort to simpler models to describe relationships; however, as shown in the following example, this can produce misleading, maybe contrary, results.

Look back at the results of the decision tree in Figure 1.7. You might find it easy to conclude that the average purchase increases directly with the income level of the purchaser. This relationship is dramatically illustrated in the first branch of the decision tree. Average purchases increase from about $220 for those consumers whose incomes are $74,900 per year or less, to $270 for those consumers whose incomes are more than $74,900 annually. A better and more thorough understanding of this relationship comes from a closer examination of the various antecedents and intervening factors that could influence this relationship.

The term *antecedent* refers to factors or effects that are at the base of a chain of events or relationships, just as planting a seed can be an antecedent to measuring stem growth. An intervening factor comes between the ordering established by the other factors and outcome (for example, earth and water can serve as intermediate sprouting media to observe the effect of the planted seed on stem growth). Intervening factors can interact with antecedents or other intervening factors to produce an interactive effect. Interactive effects are an important dimension of discussions about decision trees and are explained more fully later. Decision trees show both main effects and interactive effects. For example, in Figure 1.7, the first level (branch) of the decision tree shows the main effect of income on purchases. The second level, under income, shows the interactive effect of income by number of purchases in the sales category of juvenile purchases.

Figure 2.1 displays a classic relationship observed between **X** and **Z**. **X** can represent any number of situations, events, states, or factors, usually captured on a data record. The same is true for **Z**. Antecedents, shown as **A** in Figure 2.2, include a variety of situations, events, states, or factors that precede **X** (conceptually or temporally), and **I** illustrates a variety of situations, events, states, or factors that could intervene between **X** and **Z**. Decision trees enable you to quickly explore your hypotheses about these relationships and to scan the data set for antecedents and intervening factors that might help you better understand the relationship between income level and amount purchased.

Figure 2.1: Illustration of Direction of Relationship

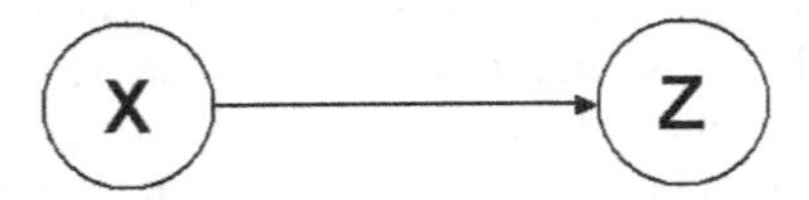

You might ask, "Does the relationship between income level and purchase amount depend on the gender of the customer?" (This question asks for an antecedent that might shed light on the relationship.) Or you might ask, "Does the relationship between income level and purchase amount depend on the number of average shopping visits in a year, or does it depend on the most recent purchase?" (This question asks for an intervening factor that could enhance your understanding of the relationship.) The results of looking at these two questions are illustrated in Figures 2.3 and 2.4 in subsequent sections.

Figure 2.2: Illustration of Antecedents and Intervening Factors

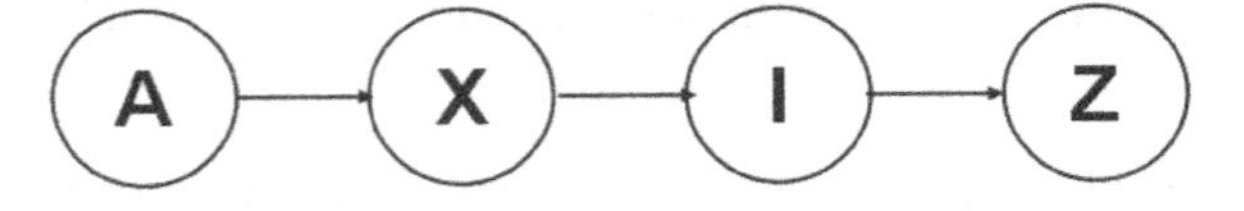

Antecedents

Figure 2.3 provides a concrete example of how an antecedent (in this case, gender) can affect the relationship between two other variables (income level and average purchase).

Figure 2.3: Illustration of the Effect of Other Factors

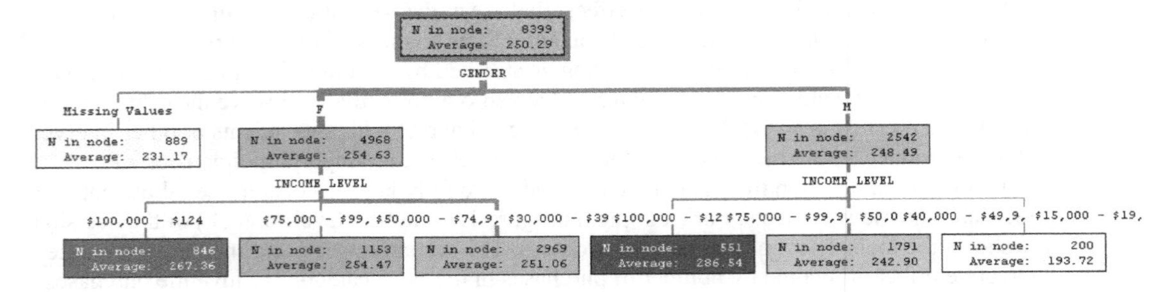

In Figure 2.3, the general form of the relationship confirms that females spend more, on average, than males, and spending increases with income level for both males and females. However, there is an anomaly in the spending of the high-income males; the males with an annual income of $100,000+ actually outspend the same category of females—$286 versus $267. One interpretation of this effect is that the very best customers (in terms of purchase amount) are not high-income females, they are high-income males. This shows how decision trees can be used to test the effects of antecedents on the form of a relationship.

Intervening Factors

The decision tree in Figure 2.4 shows the effect of the intervening factor—latency—on the form of the relationship between income level and purchase amount. The term *latency* is borrowed from physics to describe the period of time that one component in a system is waiting for another component. In this case, latency refers to the period of time when the customer is outside the purchase cycle. Generally, the greater the latency (the time since the last purchase), the lower the average purchase amount. This suggests that high-spending customers are also high-value customers.

Figure 2.4: Illustration of the Effect of Intervening Factors

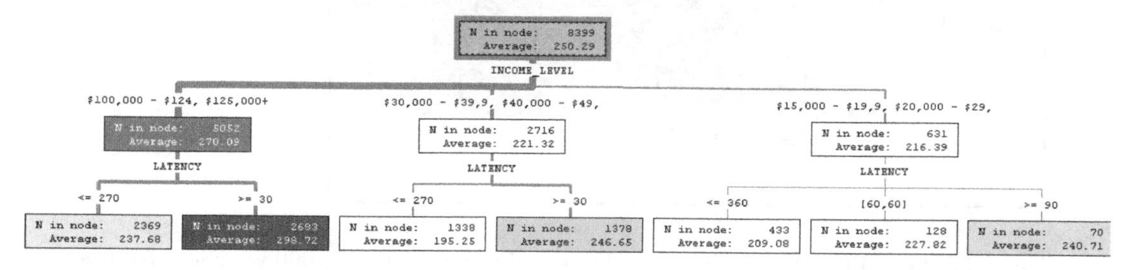

An anomaly is revealed in the decision tree in the low-income group; among the 631 people included in the survey from low-income groups (incomes of $30,000 per year or less), the amount of purchase actually increases with latency (purchasers with latency in the >=90-day range out-spent those in the 60-day range). There are several interpretations of this phenomenon; for example, low-income customers may save up money to make planned-for purchases.

The important point to note is that intervening factors can mediate interrelationships between input variables, and decision trees provide a flexible method of examining how these effects can be accommodated in the interpretation and extraction of marketing knowledge.

A Classic Study and Illustration of the Need to Understand Context

Antecedents and intervening factors can have an important effect on the form of a relationship. Many documented cases show that this effect is substantial and might involve a complete reversal in the direction of a relationship (e.g., from positive to negative), and can be both surprising and counterintuitive. A classic example is illustrated in the article "On Simpson's Paradox and the Sure-Thing Principle," in the *Journal of the American Statistical Association* (Blyth 1972). To understand the scenario presented in this article, assume that you are a marketing manager for a software development/publishing company and that you are evaluating the effects of various promotional programs on long-term software retention. In Figure 2.5, you can see that the results to date have been particularly discouraging.[1]

Figure 2.5: Illustration of Relationship Reversals—Baseline

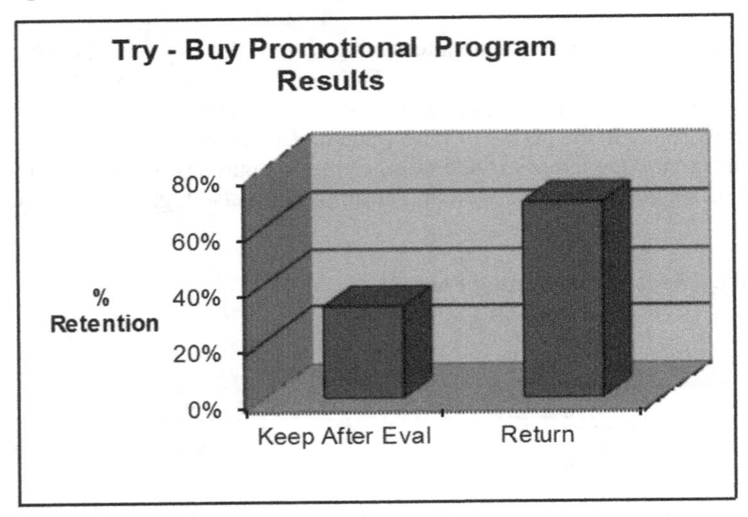

Figure 2.5 shows that a randomly selected group of respondents—11,000 were selected from advertisement responders and 11,000 were selected from information request responders—have a poor overall product retention (they buy the product after an evaluation period) of only 32%. What is even more disturbing is that it was assumed that the information request responders would have a higher product retention because, presumably, these responders were better qualified than the responders to the general advertisement. The results on the source of the response are shown in Figure 2.6.

Figure 2.6: Illustration of the Effect of Third Variables

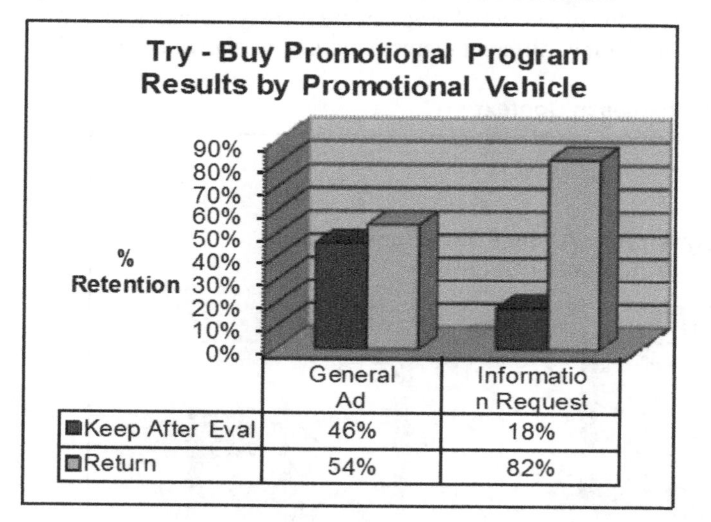

	General Ad	Information Request
Keep After Eval	46%	18%
Return	54%	82%

The marketing model assumed that, although it was more expensive to generate leads from articles and infomercials, these leads would be more likely to result in better-qualified consumers than leads from general advertising, and that these leads would, in turn, have a higher retention rate.

The results presented in Figure 2.6 demonstrate that this marketing model was completely wrong…or was it? Are there other factors present and unaccounted for that would confirm the marketing model and perhaps indicate a successful program? In other words, are there other variables that capture contextual effects that need to be looked at to more accurately understand the relationship between retention and promotion?

The Effect of Context

So far, the results have been presented without considering all of the effects of possible predisposing or intervening factors in the presentation. One such factor—customer segment—has been excluded from the current analysis. Segment membership is recognized as an important component in the overall marketing program. Because of its importance, all customers are scored on a segmentation framework that was developed to chart the value of customers. As a result, customers are managed better and new customers can graduate to higher levels of customer value.

Segmentation makes a major distinction between the software's general users (generic) and higher-value power users. When the results of the promotional program are displayed, taking these two critical segments into account, a considerably different picture emerges, as shown in Figure 2.7.

Figure 2.7: Illustration of Relationships in Context

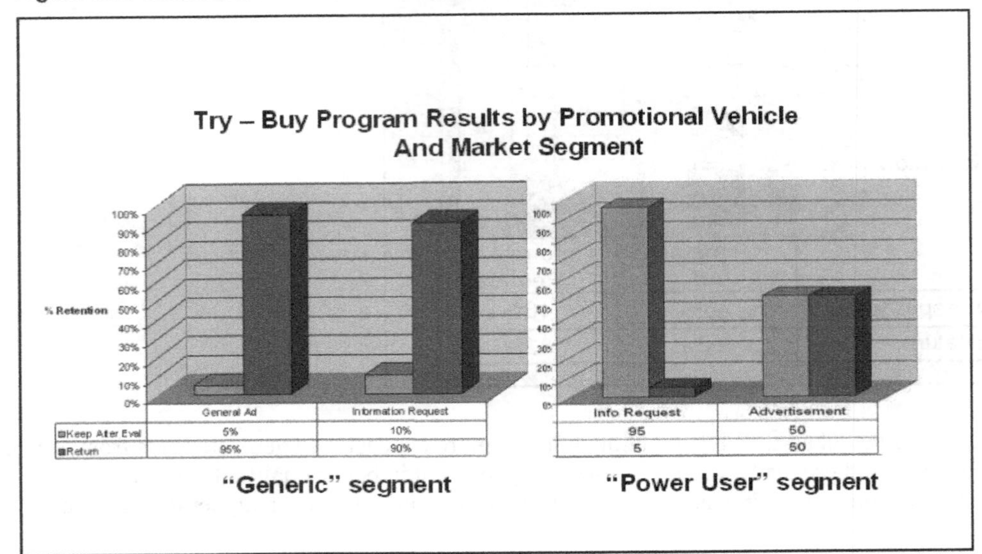

When results are presented with the important customer segments included, a different view is provided; in both customer segments, the information request promotional vehicle outperforms the general advertisement. In both customer segments, responders who were selected for the evaluation via the information request were about twice as likely to keep the software (10% versus 5% and 95% versus 50%).

How Do Misleading Results Appear?

How do the kinds of astonishing reversals of results, such as what happened in the "sure-thing principle" (Blythe 1972), occur? How can decision trees be used to ensure the discovery and presentation of valid results? The decision tree could show some of the drivers of these reversal results. In Figure 2.8 the information request vehicle appears to confirm the original assumption: advertisements are a better source of renewed business.

Figure 2.8: Illustration of Advertisement versus Information Request Promotion

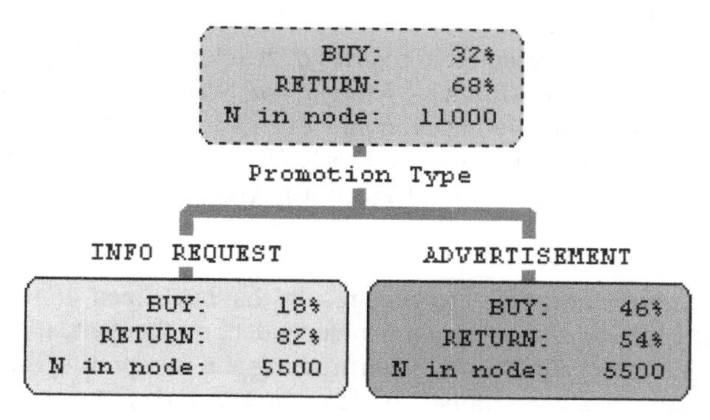

If you look at the full decision tree in Figure 2.9, however, a different picture emerges. In the favored customer segment power users, the effect of information requests as a source of renewed business is very strong. Clearly, a decision tree application that is capable of sifting through the various interactions (combinations of antecedents and intervening factors that can influence the interpretation of relationships) would be useful.

Figure 2.9: Illustration of Full Decision Tree

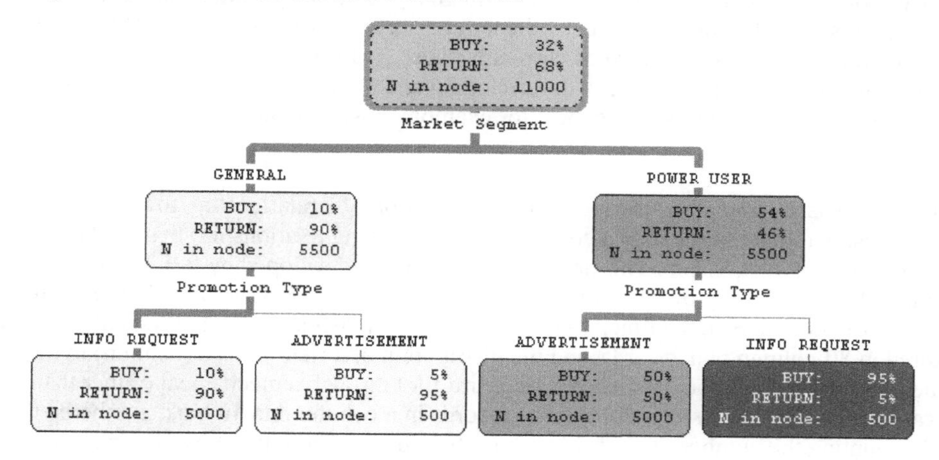

Automatic Interaction Detection

It is precisely this motivation that drove the results presented in a series of articles authored by the primary developers of decision tree software in North America—Morgan and Sonquist. They were economists working with economic indicators and predicting economic events at the Institute for Social Research at the University of Michigan in the late 1960s and early 1970s when they began to write a software program called AID (Automatic Interaction Detection). AID became the first decision tree program in North America.

Along with many other researchers at the time, they noted the same results that have been shown in previous examples—the form of a specific relationship is very much dependent on the context of the relationship and on the influence of other relevant factors in constructing and interpreting the relationship. They documented these observations in a seminal article in the *Journal of the American Statistical Association* (Morgan and Sonquist 1963). In this article, they suggested the use of decision trees to search through the many factors that can influence a relationship to ensure that the final results that are presented are accurate. The suggested approach evolved from a method of tabular analysis—a precursor to current-day multidimensional (OLAP) cubes—that was popular at the time. Although they were working within a regression framework, it is noteworthy that regression techniques, at that time, were in their infancy due to the limitations of computers. Even though Morgan and Sonquist were working with numeric data and regression models, they suggested an approach that was built on a style of contingency table analysis that had been developed by social scientists working on social theory and survey research analysis. This places this early development in the same context as current business analysts who might use regression techniques, but who are more comfortable with developing and presenting results that are based on table views drawn from multidimensional cubes (for example, business intelligence and business analytics).

This style of analysis is a systematic attempt, in the examination of a relationship, to identify a preceding relationship (sometimes called a controlling or specifying relationship) that could change the nature or form of the relationship. This analysis approach was developed by P.F. Lazarsfeld and M. Rosenberg and was originally discussed in *The Language of Social Research* (1955).[2] Because of the limitations of computers at the time, it was common to conduct a tabular analysis of data. Data was stored in 80-column punch cards, and the distribution of a field of data (column) in the card could be found by passing the data records (the card file) through a mechanical (rather than electronic) card sorter. This card sorter split up the records in a column into 10 bins, numbered 0 through 9. By counting the number of cards in each of the bins for a column, it was possible to derive the distribution of values for a field of information in the data set.

By applying this technique, it was also possible to successively partition the target of the analysis—say, income or dollars spent—according to the other fields in the analysis—for example, age of the study participants, gender, place of residence, and so on. The card sorter approach enabled the researchers to explore various subcategorizations of the target by looking at the results of various age-gender groupings and various age-gender-residence groupings, and so on. This approach produced the characteristic decision tree display that is now so familiar. Until the arrival of digital computers, this approach served as more computationally accessible compared to regression.

This approach was eventually adopted and embedded in the code that resulted in the development of the AID software program. As decision trees evolved, the goals of the approach expanded to handle both continuous and categorical table cell entries and multi-way branches. Statistical tests and validation approaches were later developed to assure the integrity of the decision tree.

In Morgan and Sonquist's approach, the type of intervening effect shown in the previous marketing example is due to an interaction between customer segment and the effect of the promotional program versus retention. The overall effect is negative as you move from information requests to advertisements, with respect to keep versus return. Yet, the interaction displays subregions of the relationship that are dominated by the predisposing factor of customer segment. Within a customer segment, the relationship is positive. In a term that was introduced by Lazarsfeld and Rosenberg, this is an example of a "controlling" relationship.

The concept of an interaction effect—or controlling relationship—is common in many modeling situations. As originally pointed out by Morgan and Sonquist, an interaction can obscure a strong relationship. In their article, they produced an example (Figure 2.10) where there is a relationship between savings and income—but only for the self-employed. There is an interaction between employment status and rate of savings. In Figure 2.10, the effect of employment status "specifies" the relationship between savings and income; it shows a more specific relationship among the various income categories.

Figure 2.10: Illustration of an Interaction Effect

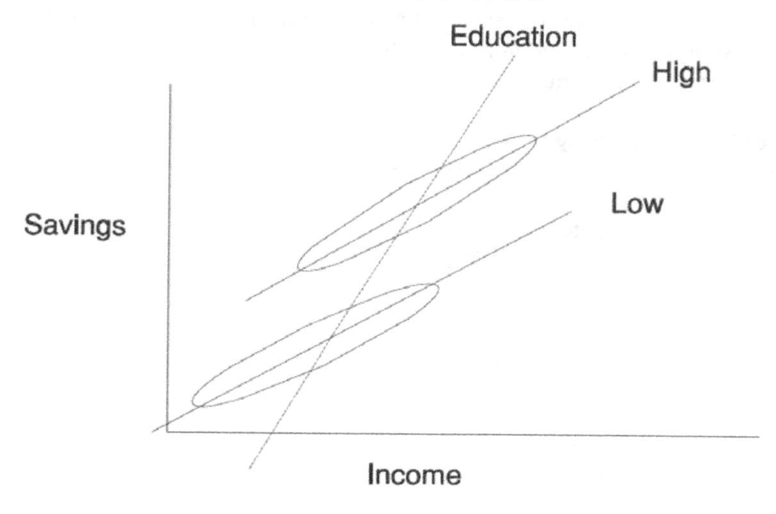

Morgan and Sonquist proposed the use of this approach as an improvement over standard regression models. Thus, the automatic discovery of this kind of interactive relationship could be used to either grow a decision tree as an alternative to regression, or as a means to introduce interaction terms in the regression equations. The interaction term is used to segment the regression equation into two slopes; one slope captures the relationship between savings and income for the self-employed, and the other slope captures the relationship between savings and income for others.

Morgan and Sonquist noted that the decision tree approach provided an explanation of about two-thirds of the variability in the savings-income relationship, while the regression approach, even with interaction terms in the equations, accounted for only 36% of the variability. So, although decision tree results can be used to improve regression equations, these improvements might not perform at the same level as the original decision tree. This observation and the resulting inquiry—and exposition of the relative merits of regression versus decision trees—prompted a lively discussion that continues to this day.

Morgan and Sonquist discovered and published an extremely important consideration regarding the complementarity and substitutability of regression and decision tree approaches: it is normal for decision trees to perform well with strong categorical, nonlinear effects. Even when these effects are used to enhance the regression equation, the regression results can still be inferior to the decision tree results. However, decision trees are inefficient at packaging the predictive effects of generally linear relationships and, in this situation, regression tends to perform better (and yield more economical models).[3]

Morgan and Sonquist discussed using the AID decision tree approach in dealing with another common problem with regression equations—multicollinearity. In multicollinearity, the relationships between the predictive terms in an equation obscure their effect on the target. This problem is shown in Figure 2.11. An appropriate remedy for multicollinearity is to respecify the regression equation (in this example, you would introduce a high-low savings term in the regression model to force separate slope estimates).

Figure 2.11: Illustration of Multicollinearity

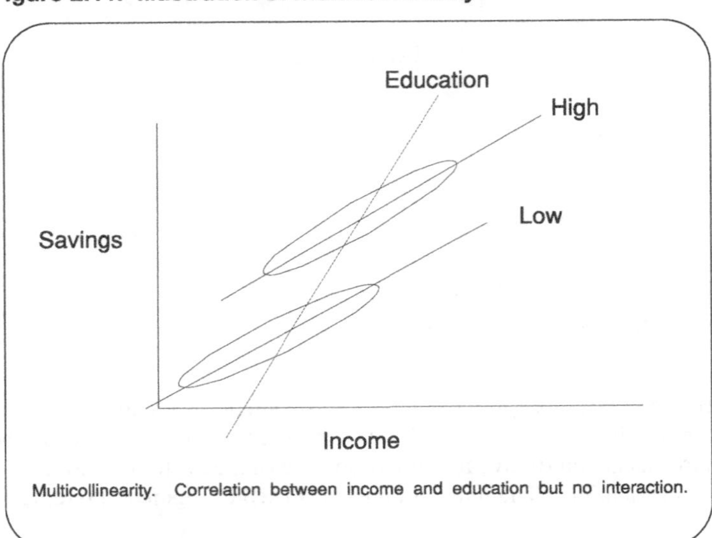

Multicollinearity. Correlation between income and education but no interaction.

Figure 2.12 shows how a regression model could be masked by a combination of both interactive and multicollinear effects. In this situation, decision trees would be immune to the model-defeating characteristics of these effects, and would be a useful tool in identifying terms for the regression equation to help the models perform better (and yield more interpretable results).

Figure 2.12: Illustration of an Interaction with Multicollinearity

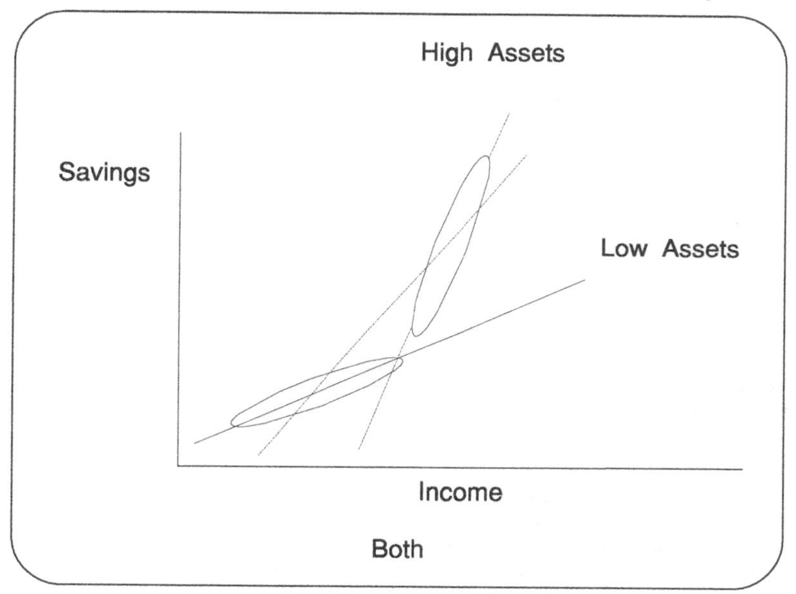

The power and utility of decision tree methods and of the original AID software program is that both decision trees and AID addressed the problem of hidden relationships. This type of analysis technique proved to be very popular. The publication of Morgan and Sonquist's results, coupled with the availability of decision tree software (both in the United States and Europe), led to the development of decision trees as a stand-alone analysis technique. For many analysts, including statistical analysts, it became simpler and just as effective to use decision trees alone, which avoided the requirement of respecifying the regression equation. This decision tree popularity coincided with the growing power of computers and the ability of statistical analysts to move out of a tabular analysis framework and into a regression framework, or into a regression-augmented-with-AID framework.

It should be noted that interpretability is sometimes overlooked as a desirable model feature in its own right, particularly as quantitative methods continue their migration into more general areas of business use. A decision tree display is often superior to the purely numerical display of the regression model because general users can recognize the qualitative and visual characteristics of a decision tree. By the same token, a general user can more easily recognize a regression-line display of a regression equation.

Figure 2.13: Illustration of Numerical, Regression, and Decision Tree Displays

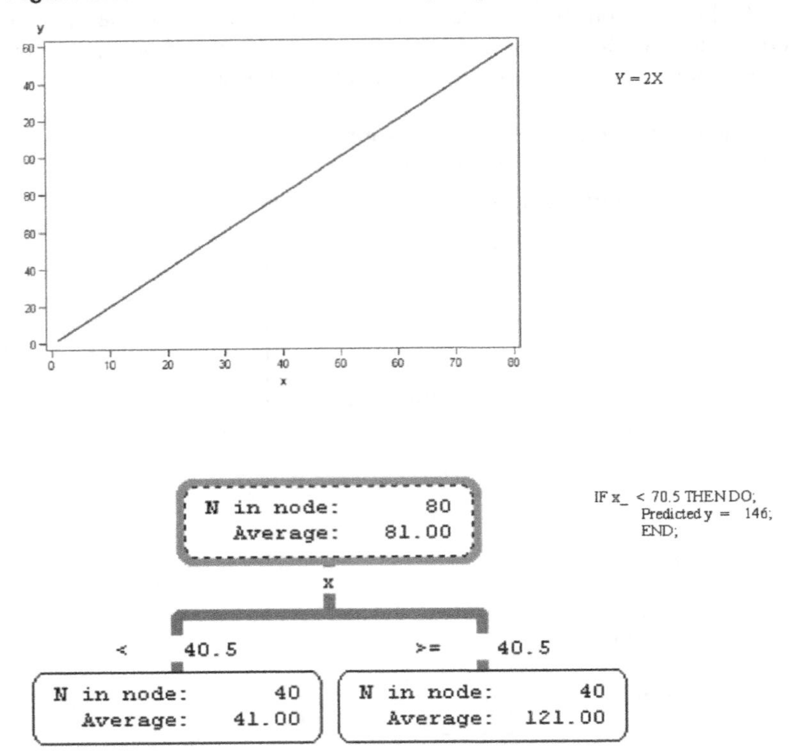

The Role of Validation and Statistics in Growing Decision Trees

Although it had several benefits and utilities, the AID decision tree approach developed by Morgan and Sonquist turned out to have major problems. The first problem was caused by the strength of the numerical-searching algorithm. Essentially, because the algorithm looks through so many potential groupings of values, it is more likely to find groupings that are actually anomalies in the data. Further, the successive partitioning of the data set into bins that form the nodes of the branches of the decision tree quickly exhausts the number of observations that are included in lower levels of the decision tree. As a result, the successively lower levels of the decision tree are based on an increasing number of assumptions about the splits that are used to form the branches. Also, due to the recursive nature of the decision tree algorithm, fewer data records and associated data points are used to identify the specific leaves or nodes that are formed by the branch splits or partitions.

Both problems—fitting anomalous relationships and fitting relationships with limited data—meant that it was not always possible to believe the efficacy of the branches that were identified by AID.

This was pointed out as early as 1972 by Einhorn in an article that demonstrated that AID could form branches that reflected idiosyncrasies in the data, rather than reflecting effects in the population that the data represented. He pointed out that branches were formed based on a statistic that tried to minimize the variance within nodes and maximize the difference between nodes. The advantage of AID—looking through data and identifying any branch or split that could be used as an interaction—meant that many splits were formed and examined. This led to a "data-dredging" effect, where inputs formed branches with numerical values (branch partition values) that showed overstated results. These results were no more than artifacts of chance. The overstated results can be produced by the intensive computation to identify combinations of values that can be used to form branches of a decision tree. When many combinations are examined numerically, it is usual to identify combinations that favor a particular view of the data that reflects the idiosyncrasies of the data, not the characteristics of the universe that the data was drawn from. This is the result even if the sample data is an accurate reflection of the universe that the sample data was drawn from.

An additional problem with AID was its tendency to find branches in inputs with large numbers of values, to the exclusion of branches in inputs with smaller numbers of values. This, too, was an artifact of chance and computation. For example, when looking for a binary split in a range of 100 values, AID would form a split for 1 versus 2+, 1-2 versus 3+, and so on. This process increases the chance that there will be at least one split along this range of 100 values that shows differences in a given target. In contrast, when exploring the relationship between gender, there is only one split possible—male versus female. If this branch does not produce a strong effect, then the algorithm will examine another input. There is less opportunity for chance to produce an effect (as there was with 100 values) and, consequently, there is less opportunity for fields with relatively few values to enter into the model when compared to fields with relatively more values.

Overall, three different kinds of problems were noted with AID:

- "untrue" relationships (e.g., showing structure in random data)
- biased selection of inputs or predictors
- an inability to know when to stop growing branches, and forming splits at lower extremities of the decision tree where few data records were actually available

Because of its growing popularity and the utility of AID as an analysis tool, remedies to the problems were proposed. Remedies include using statistical tests to test the efficacy of a branch that is grown. For example, if a branch shows a difference between males and females with respect to an outcome, is this difference significant from a statistical point of view? Another remedy involves using validating data to test any branch that is formed for reproducibility. Hold-out or validation data is typically formed by drawing a random sample from a data set before the data set is introduced into an analysis. The hold-out data is used to test any relationships that are formed with the original data (minus the hold-out data). Because the original data is used to form the relationship, it is sometimes called the "learning" or "training" data (because the algorithm "learns" the relationship).

One of the first remedies for addressing the problems with AID was proposed by Kass (1975). Kass suggested the use of statistical tests and Bonferroni adjustments. Bonferroni adjustments are named after the statistician who suggested that the level of statistical confidence of a statistical test be adjusted to account for the number of tests or trials that were used in producing the test. This provided a means to place inputs with 100 values or 2 values on the same footing. And it overcame the AID tendency toward the biased selection of predictors. These statistical tests also tested the reliability of branches formed in the AID decision tree, including the branches at lower levels of the decision tree.

Kass's approach was called CHAID. This stood for "chi-square AID". In conjunction with Hawkins (another statistician), Kass developed another approach, called XAID, which also used statistical tests, but worked with continuous targets (Hawkins and Kass 1982).

Remedies based on a validation approach were soon proposed by Breiman, Friedman, Olshen, and Stone (1984). Whereas the Kass approach used classical statistical theory to address the shortcomings of AID, the Breiman et al. approach relied on validation techniques to improve upon AID. Breiman et al. also introduced a number of new features. Their approach was called Classification and Regression Trees (CRT) and was published in a book of the same name.

The Application of Statistical Knowledge to Growing Decision Trees

Solutions based on Kass's and Hawkins's methods began to appear in the late 1970s. The CHAID method works with a categorical response or target. The XAID method works with a continuous (or numeric) response or target. The general approach of their methods—referred to as CHAID analysis—allows for the development of decision trees with both categorical and numeric targets. The inputs to the analysis are used to form the attributes of the decision tree. The inputs, like the target, can be categorical or numeric. Although branches are formed as categories, Kass provided a method of dealing with numeric data that is at ordinal or interval levels of measurement. (It is usually possible to compress ordinal or interval data into a more restricted range of categorical values.)

Significance Tests

Statistical tests that are used in the CHAID analysis approach:

1. CHAID methods use a test of similarity to determine whether individual values of an input should be combined. For example, if two age values have the same response value (from a statistical point of view, they are indistinguishable), then they are combined.
2. After similar values for an input have been combined according to the previous rule, tests of significance are used to select whether inputs are significant descriptors of target values and, if so, what their strengths are relative to other inputs.

The approach developed by Kass addresses all the problems in the AID approach:

- A statistical test is used to ensure that only relationships that are significantly different from random effects are identified.
- Statistical adjustments address the biased selection of variables as candidates for the branch partitions.
- Tree growth is terminated when the branch that is produced fails the test of significance.

Kass introduced another innovation in the development of the form of the decision tree by describing how to form multi-way splits in the branches of the decision tree (as opposed to the simple binary or 2-way splits that form the AID decision tree). This multi-way splitting emerged as a result of Kass developing what he described as a merging-and-splitting heuristic in the construction of the branches of the decision tree.

The Role of Statistics in CHAID

CHAID relies on a traditional statistical test of significance to form the group boundaries that determine the values of the inputs that form the branches of the decision tree. Traditionally, the test of significance is constructed around the null hypothesis. When comparing the distributions of two or more groups in a data set, the statistician gathers numeric evidence to characterize the two or more groups and then poses the question, "Are there differences in magnitude among the groups so great that the null hypothesis of no differences can be rejected as not tenable?"

In practice, as greater magnitudes of differences are observed among groups, the statistician has more confidence in the structure and form of the relationship.

Confidence				
Extremely Good	**Good**	**Pretty Good**	**Not So Good**	**Extremely Weak**
.001	.01	.05	.10	.15

This test of significance determines which values are combined. The values are used as the various inputs that are considered as splitting criteria in the construction of the decision tree. To elaborate, a test of significance determines whether two values are the same with respect to their relationship to the target. If their values are the same, then they are combined. If their values are different, then they are separate branches on the decision tree. The significance test is illustrated in Figure 2.14.

The degree of separation between two groups can be used as a test of the difference between two groups. The larger the separation, the stronger the relationship and, consequently, the greater the statistical confidence in the relationship. Because any two nodes on the branch of a decision tree can be seen as two groups, the internode separation can be tested with a test of significance. Multi-node tests can be used just as multigroup tests are used.

Figure 2.14: Illustration of Tests of Significance

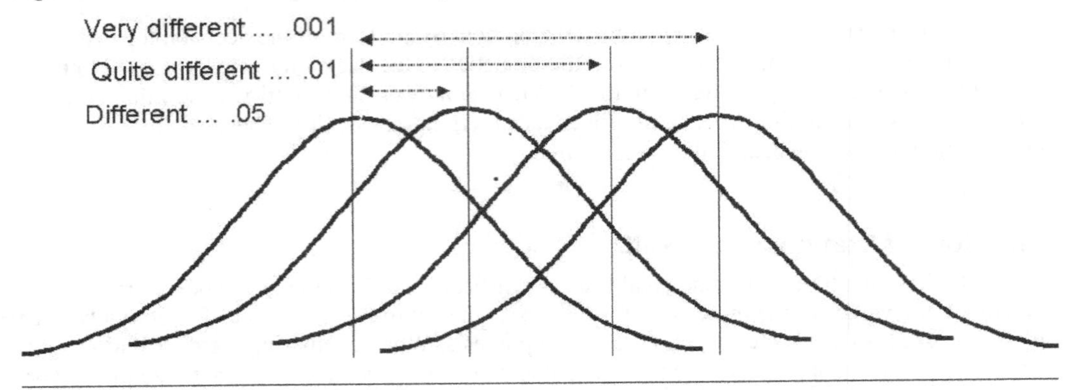

The second way that CHAID methods use statistics is to judge which relationships are strong enough to use in building the model. Once the values of a given input to the CHAID decision tree are combined through the merge-and-split method, then the resulting table can be set aside for subsequent evaluation. The process of combining field values for each input in the decision tree continues using the merge-and-split heuristic until, finally, all inputs have had their values combined.

Figure 2.15: Significance as a Function of Distribution Separations

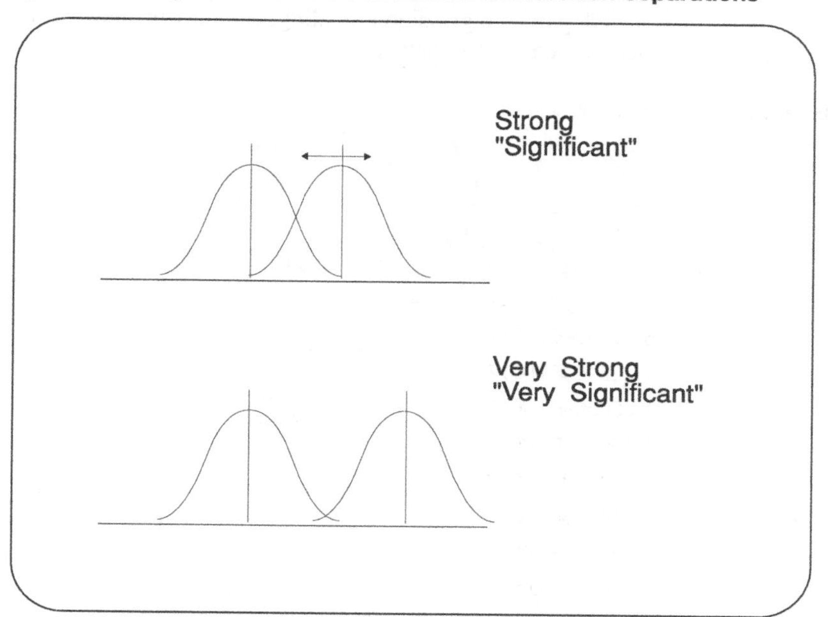

After the values of the inputs have been combined, you can look at each of the inputs and determine the overall statistical relationship between a given input, its associated branches, and the target to be predicted.

Kass proposed the use of statistical adjustments—referred to as Bonferroni adjustments—to eliminate the side effects of data-dredging. The level of statistical significance used to assess the identification of branches on the tree is adjusted. This adjustment factors in the number of tests that were conducted in identifying the relationship.

After Bonferroni adjustments have been applied, alternative partitions can be presented to the decision tree display, and the most appropriate input can be selected as the splitting criterion. In the absence of any other criterion, the input that is selected is the input with the highest Bonferroni-adjusted level of significance. Although this could be the best way to grow a decision tree from a predictive point of view, it might not be the best way to show the nature and sequence of relationships that characterize a given target. It is preferable for the analyst to grow the decision tree so that it supports the conceptual model that is being used to describe the target. So, when examining a list of alternative branches at a given level of the tree—all branches being significant—the analyst might choose the branch that best fits the conceptual model. This type of choice (from the SAS Enteprise Miner interface) is shown in Figure 2.16. In a banking application, a number of variables—such as IRA Balance, Age, and so on—are potential inputs. Although the variable IRA Balance has the highest splitting criterion value (9.65622), there are many other variables that could be selected to grow the decision tree. Any of the variables could be used as a splitting criterion because all of them are significant from a statistical point of view. Selecting branches in a particular sequence has analogies in regression modeling, whereby the entry sequence of terms in a regression equation is determined by the analyst.

Figure 2.16: SAS Enterprise Miner Display

The measure of significance **-Log(p)** is a transformation of the normal method of displaying significance. This transformation is shown in the Glossary at the back of this book.

Validation to Determine Tree Size and Quality

While Kass was improving the operation of AID through tests of significance, parallel research and development was going on to validate data in the construction of decision trees. The results of this research and development were published by Breiman, Friedman, Olshen, and Stone (1984). The data validation approach developed by Breiman et al. was called Classification and Regression Trees (CRT).

CRT closely follows the original AID goal, but with improvement through the application of validation and cross-validation. In CRT, it is easy to determine where there is overfitting; as the decision tree is being developed, construct an algorithm to verify the reproducibility of the decision tree structure using hold-out or validation data. After a decision tree or a branch of a decision tree is grown, reproduce the growth in the hold-out or validation data. If the validation results deviate from the training results, then the decision tree is not stable. Typically, the top level of the decision tree is readily reproduced; however, at lower levels of the decision tree, training results and validation results tend to deviate. And, at some level, the deviation is too severe to retain the form of the decision tree.

Breiman et al. found that it was not necessary to have hold-out or validation data to implement this grow-and-compare method. A cross-validation method can be used by resampling the training data that is used to grow the decision tree. This resampled data can also be used as a reference point—relative to the original or training data—to check and verify the accuracy and reproducibility of the tree as it is being grown.

CRT can include complexity (parsimony) to tune the size of the decision tree. With CRT, lower branches are penalized in the validation, which makes it harder to grow bigger trees that pass all the validation tests. In addition, it is possible to use prior probabilities to tune the size and shape of the decision tree. Here, the validation test is adjusted to reflect a distribution of the validation statistics that are calculated, so that the validation test is calculated on the basis of this distribution (rather than on the raw distribution, as reflected in the hold-out validation data).

What Is Validation?

Validation is a method of verifying the integrity (reproducibility) of a statistical model. Validation works by setting aside test data (typically 30%, selected randomly) that is from the original (training) data set used to develop the statistical model. This test data is subsequently used to test the performance of the model that is developed with the original (training) data. This form of validation is an alternative to resubstitution. Resubstitution uses data twice—once to grow the model and then again to test it. Simple methods of resubstitution overestimate the model's integrity.

Figure 2.17: Illustration of Validating Training and Test Data

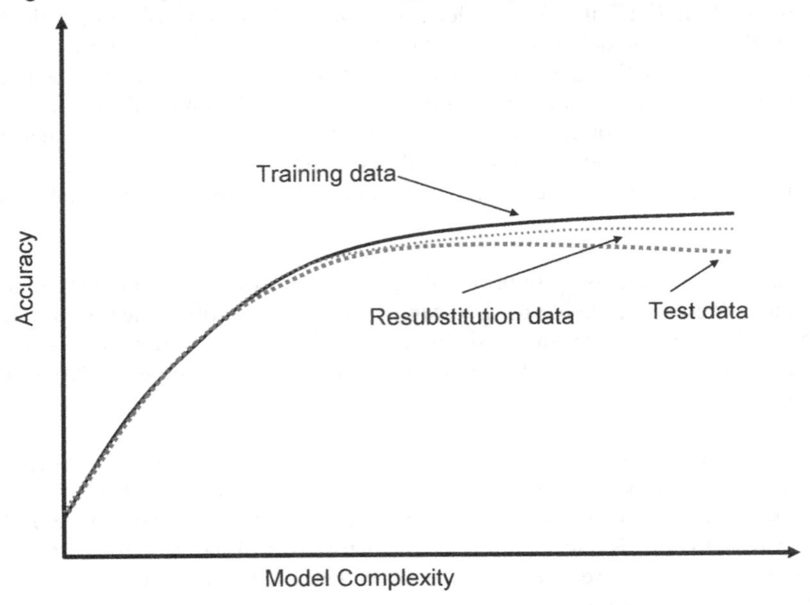

Regardless of the method used to validate the decision tree, you can assess the stability of the decision tree by comparing branch distributions of the target in the training and test data; if the test decision tree produces results that match the training decision tree, then there is confidence that the branches are reproducible and accurate.

In practice, training and test decision trees are built branch by branch. Comparisons between training and test decision trees are made with each successive branch that is built. Comparisons are made on the basis of deviations in the target values in the respective decision trees. If the target value is categorical, then the modal (most common) category predictions in the training decision tree are compared to the test tree. For interval targets, training mean values, as well as the variability between training and test samples, can be compared for similarity. Comparisons are made on the basis of error rates.

At some point when growing branches, the error rates between training and test samples begin to diverge. As error rates climb the decision tree, the stability and reproducibility of the respective training and test trees at the lower branches deteriorate. When deterioration begins, it is time to stop growing the tree and select a subtree consisting of the higher branches that are more stable.

Figure 2.18: Illustration of the Grow-and-Prune Strategy

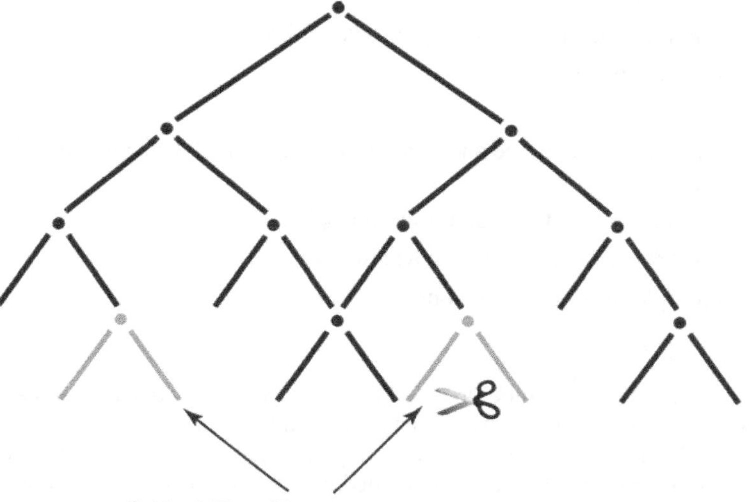

The development of CRT laid the groundwork for these and many other validation approaches and provided tractable methods to grow reliable and accurate decision trees (what Breiman et al. called "honest" decision trees at the time). CRT solved the problems with the AID approach and proved as powerful a technique as the CHAID and XAID approaches developed by Kass.

The full methodology for growing and pruning branches in CRT includes the following:

- For a continuous response field, both least squares and least absolute deviation measures can be used. Deviations between training and test measures can assess when the error rate has reached a point to justify pruning the subtree below the error-calculation point.
- For a categorical-dependent response field, it is possible to use either the Gini diversity measure or Twoing criteria.
- Ordered Twoing is a criterion for spitting ordinal target fields.
- Calculating misclassification costs of smaller decision trees is possible.
- Selecting the decision tree with the lowest or near-lowest cost is an option.
- Costs can be adjusted.
- Picking the smallest decision tree within one standard error of the lowest cost decision tree is an option.

In addition to a validated decision tree structure, CRT provided other extensions to AID:

- works with both continuous and categorical response variables
- handles missing values by imputation
- employs surrogate splits
- grows a larger-than-optimal decision tree and then prunes it to a final decision tree using a variety of pruning rules
- considers misclassification costs in the desirability of a split
- uses cost-complexity rules in the desirability of a split
- splits on linear and multiple linear combinations
- does subsampling with large data sets

Like AID, CRT employs a binary splitting methodology, which produces binary decision trees. CRT does not use the statistical hypothesis testing approach proposed by Kass, and CRT relies on the empirical properties of a validation or resampled data set to guard against overfit. Breiman et al. did not embrace the kind of merge-and-split heuristic developed by Kass to grow multi-way splits, so multi-way splits are not included in the CRT approach.

Pruning

The role of validation and pruning can be described using a decision tree run against a data set of banking transactions. The data set contains credit score as a target variable, and a number of inputs, including customer demographics, banking attributes (such as accounts used), and behavioral data such as transaction timing, counts, and monetary value. Figure 2.19 illustrates how a decision tree can grow with or without validation on the same data set (here, banking customers). As shown in the figure, a different approach to validation can produce dramatically different results.

Figure 2.19: Illustration of a Pruning Scenario

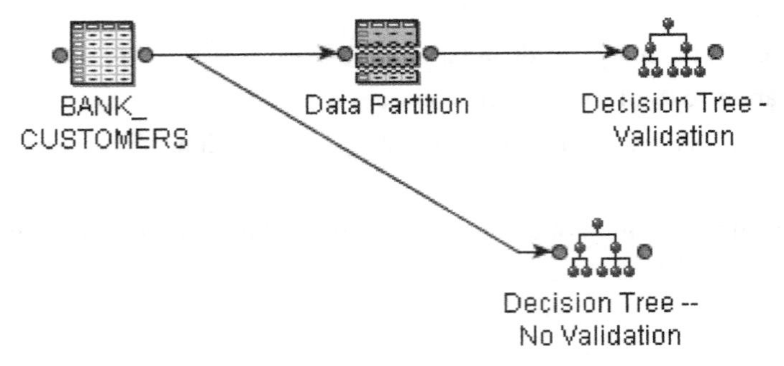

With a continuous target (such as credit score), a typical measure of decision tree model accuracy is to use an average squared error comparison between the training data set and the validation. The results are displayed in an iteration chart that shows the leaf-to-leaf values of average square error for the training and validation data sets respectively.

Figure 2.20: Illustration of an Iteration Chart

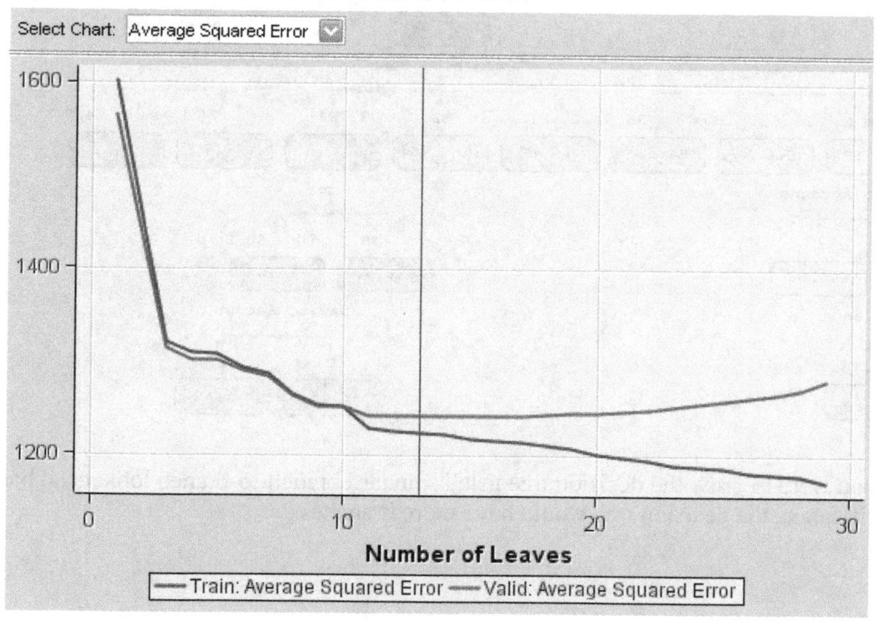

The training and validation data set decision trees yield a similar average squared error through the construction of 10 leaves, but they begin to diverge in the construction of leaves 11, 12, and so on. This results in the decision tree shown in Figure 2.21.

Figure 2.21: Illustration of the Effect of Pruning Decision Tree Growth

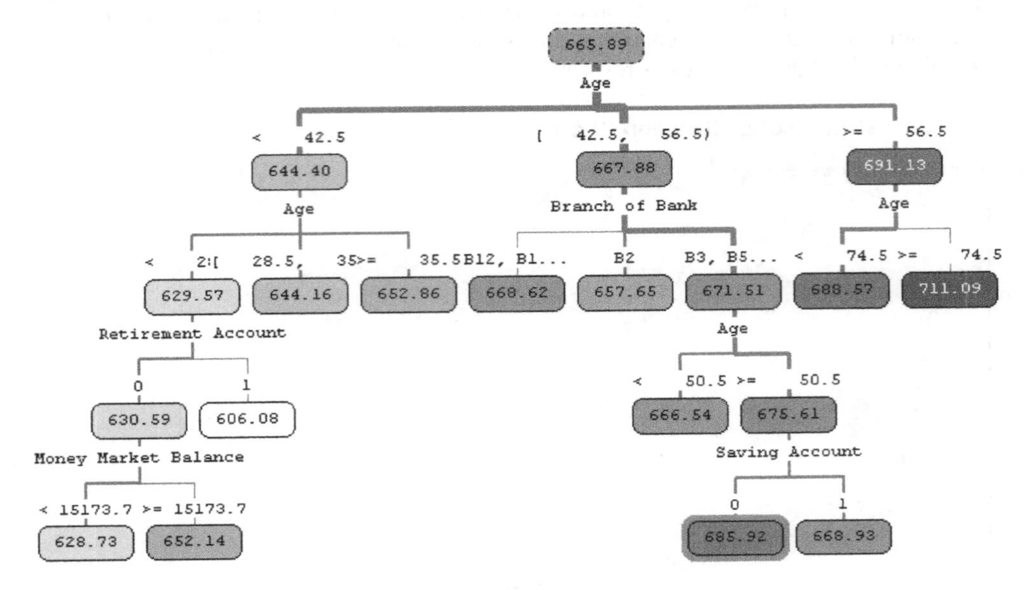

By contrast, if you were to grow the decision tree using standard branch-to-branch lookahead based on a test of significance, the decision tree would have more branches.

Figure 2.22: Tree Growth Using Significance Tests to Stop

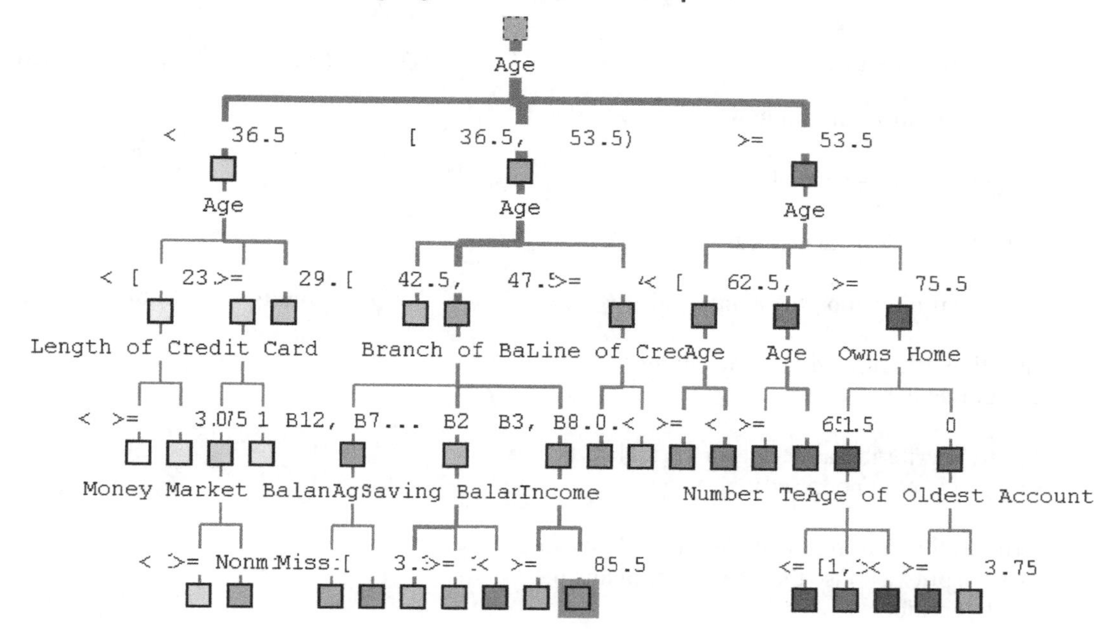

This decision tree has a total of 24 leaves, which is twice as many as what was produced using pruning as shown in Figure 2.21.

Decision trees have evolved significantly; however, early forms of decision trees laid the groundwork for many adaptations. The significant characteristics of these early decision trees are summarized in the following table.

Table 2.1: AID, CHAID, XAID, and CRT Compared

Technical Feature	AID	CHAID	XAID	CRT
Target variable - continuous	□		□	□
Target variable - ordinal				□
Target variable - categorical		□		□
Degree of branch partitioning	2	multi	multi	2
Splitting criterion adjusted for number of categories in predictor		□	□	
Splitting criterion adjusted for variable type (e.g., ordered, unordered)		□	□	
Criteria for splitting the node—for all possible 2-way splits, choose the one that explains the most variation	□			□
Criteria for splitting the node - for all possible multi-way splits, choose the one that explains the most variation		□	□	
To determine final decision tree stop: when too few observations	□	□	□	□
To determine final decision tree stop: when no more splits are significant		□	□	
Build a large decision tree and prune based on validation tests formed from test sample or resampling approach				□
Split on linear combination of predictors				□
Prior probabilities and cost function can be specified				□

The entry "Split on linear combination of predictors" formed the basis of a number of enhancements in later years, such as the QUEST algorithm (Quick, Unbiased, Efficient, Statistical Tree). These enhancements are described by Loh and Vanichsetakul (1988). Although they are statistically powerful, these enhanced splits have never been adopted on a large scale. Splits that are made on a linear combination defeat one of the primary advantages of decision trees: ease of use and ease of interpretation and comprehension.

One of the early inspirations for decision trees—the Concept Learning System (CLS)—was proposed by Hunt, Marin, and Stone (1966). It inspired a parallel development of decision trees in the areas of machine learning and artificial intelligence. The connection between statistical approaches and approaches based on pattern recognition (including machine learning) has continued through the development of decision trees. In addition, this connection has been a robust source of innovation as decision trees have developed into a mature method of data mining.

Machine Learning, Rule Induction, and Statistical Decision Trees

Machine learning is a general way of describing computer-mediated methods of learning or developing knowledge. Machine learning began as an academic discipline. It is often associated with using computers to simulate or reproduce intelligent behavior. Example application areas include robotics, speech recognition, and language understanding and translation. Machine learning has also been used to build intelligent chess-playing programs (Shapiro 1987). Machine learning and business analytics share common goals: In order to behave with intelligence, it is necessary to acquire intelligence; further, it is necessary to acquire intelligence, and even refine it, over time as circumstances change. In these circumstances, there is a strong incentive to acquire intelligence from databases, which serve as records of positive and negative outcomes. An advantage of acquiring intelligence from databases is that the acquisition process can unfold in an automatic fashion. So, if data is accumulated less automatically, and if intelligence is extracted automatically, then it is possible to build and refine knowledge in ways that are not possible manually.

The broad goals of machine learning can be roughly compared to human learning goals: through the study of and experimentation with a particular area or subject, you can learn how the area or subject operates, how to react to it, and, possibly, how to exploit it to achieve whatever purpose you have in mind.

Knowledge can be captured and expressed in many ways and forms; for example, both collections of books and collections of data contain knowledge. Because data sets are usually more structured than books, they are a desirable source of knowledge for machine learning applications.

All decision trees are collections of rules. Although decision trees appear to be visual representations, if you look underneath, you will see that decision trees are rule expressions. Thus, every branch on the decision tree has a semantic description and because of this, decision trees are natural forms of machine learning. The development of decision trees to form rules is called *rule induction* in machine learning literature. *Induction* is the process of developing general laws on the basis of an examination of particular cases.

The areas of rule induction, machine learning, and statistical decision trees are closely linked. A good discussion of these areas and some useful references are provided by Michie (1991) and McKenzie et al. (1993). Many forms of machine learning work with data in an approach that is analogous to statistical approaches, and attempt to achieve results that are comparable to statistical results. Statistical approaches are used in the aspects of science that depend upon observations to confirm or deny objective indicators of the theories and hypotheses that explain events and phenomena in academic disciplines. Physical scientists often use empirical data to confirm their theories and hypotheses (for example, the continued effect of humidity on oxidation rates in various metal composites). Sciences of human behavior use empirical data to confirm theories (for example, increases in purchases in response to a lowered rate of interest or promotional discount). The role of statistics in these examples is often to assess the importance and reliability of the rules or relations that are discovered through the examination of the data. In this respect, the goals of statistics and machine learning are so aligned that, in many cases, they are indistinguishable (at times, they are similar disciplines with separate names). This is particularly true in the field of data mining, which explores the use of generally available data sources to extract knowledge, often in the form of rules, in order to illuminate a practical or academic concern.

Rule Induction

In the early years of the academic study of intelligence, it was common to think of knowledge and thinking processes as consisting of rules and the processing of rules. Humans, for example, could be considered rule processors who make decisions based on rules that they carry in their heads. So, if the weather is cool, you put on a warm coat. Early forms of machine learning were modeled after this conceptualization of human behavior. One of the earliest forms of machine learning was based on a form of rule induction called the Concept Learning System (CLS) and was developed by Hunt, Marin, and Stone (1966). Most forms of decision trees can trace their roots back to CLS.

Forms of rule induction inspired by the CLS algorithm (and the underlying concept learning model of Hunt, Marin, and Stone) most closely resemble statistical decision trees. Here, a concept is learned by discovering rules that can classify an object. An object is classified by discovering how variations in a criterion attribute can be predicted or explained in terms of the other attributes that have been collected or measured for the object.

In most applications of rule induction, the goal is to examine a set of cases to inductively derive predictive rules that enable you to characterize a situation with accuracy and reliability. For example, if you observe that in winter, at high altitudes, the temperature is lower, then you might propose the following predictive rule:

```
IF      season is winter
AND     altitude is high
THEN    temperature is low
```

Almost all computer systems that rely on machine learning contain at least some rules, and the majority of computer systems rely on rules to accomplish most of their main functions. A rule has the following form:

```
IF      <condition>
THEN    <action>
```

A condition can be a state that is determined by the results of an equality (for example, is age equal to 30 years?) or an inequality (for example, is hair color not blonde?) relationship.

Rules can be collected from experts or extracted from an appropriate data set. For example, in a medical expert system, a medical practitioner might propose a rule such as:

```
IF      temperature-elevated
THEN    prescribe-remedy
```

In this example, the rule reflects medical knowledge (and, in this case, conventional wisdom) that an elevated temperature usually indicates that the subject is fighting off an infectious organism, such as a cold, or has a bacterial infection; therefore, the subject requires some kind of remedy.

A study discussed by Ho Tu Bao (2002) provides real-world data on meningitis that was collected at the Medical Research Institute, Tokyo Medical and Dental University, from 1979 to 1993. The database contains data of patients who suffered from meningitis and who were admitted to the department of emergency and neurology in several hospitals. A pattern discovered from this database is expressed in the form of rules:

```
IF     Poly-nuclear cell count in CFS <= 220
        AND  Risk factor = n
        AND  Loss of consciousness = positive
        AND  When nausea starts > 15
THEN   Prediction = Virus
```

Rules can be extracted from data quickly and inexpensively. If the data is structured appropriately, then the rules are not subject to human bias and can be thought to reflect objective truth. Data can be designed to quickly respond to and reflect the environment. Thus, rules extracted in this fashion are always up-to-date. In summary, rules generated by software possess many advantages:

- Because rules are extracted from data, they are objective and not prone to subjective interpretation; they are as good as the data they are extracted from and the extraction method that is used.
- Rules can be extracted automatically. Hence, they are less expensive.
- Because rules are extracted automatically, they can be produced quickly.
- Although subjective experience and domain knowledge often cannot keep up with changes in the environment (for example, new external market constraints, new technology, and so forth), data can if it is collected properly.

Improved methods of mineral exploration is one of the many uses of rule induction in the discovery of knowledge in data. This is illustrated by the work of the Geological Survey of Canada (Reddy and Bonham-Carter 1991). Reddy and Bonham-Carter have used an inductive approach to predict mineral deposits. A database contains information on the presence of a given mineral deposit. Each record in this database also contains information about the surrounding geology, gravity, magnetic vertical gradient readings, proximity to volcanic sites, and so on. By inductively examining the conditions that are associated with the presence of a mineral deposit, it is possible to formulate a rule that predicts the location of a mineral deposit. For example:

```
RULE_1
IF     geology = mafic intermediate volcanics
OR     mafic intrusives magnetic vertical gradient = 1 to 6

THEN
no deposit = 86.6%
deposit = 13.4%

RULE_2
IF     geology = mafic intermediate volcanics
OR     mafic intrusives magnetic vertical gradient = 7

THEN
no deposit = 60.3%
deposit = 39.7%
```

Both of these rules predict a higher likelihood of mineral deposit than the average of about 5% in the entire database. Information on the surrounding geology and magnetic vertical gradient readings enables the development of these predictive rules. These rules were developed using a decision tree.

Michie and Sammut (1991) have shown that not only can decision tree rules be used to examine remotely sensed data or medical records, but they can examine physical behavior to derive a set of rules for balancing a pole, controlling a satellite, or even flying a plane. An article in *AI Magazine* (Michie and Sammut 1991) described their work with a pole and cart problem, shown in Figure 2.23.

Figure 2.23: Illustration of the Motion Dynamics in the Pole and Cart Problem

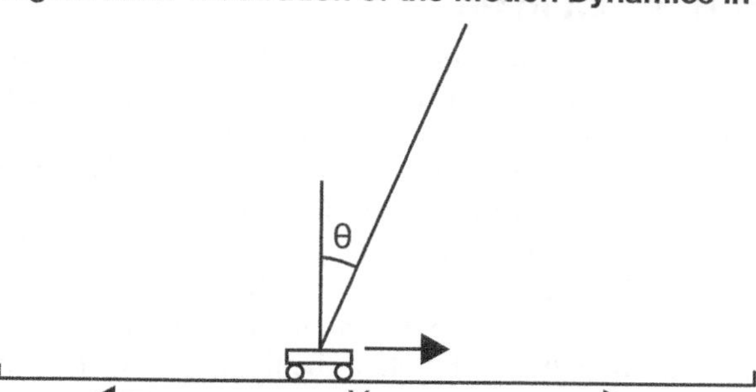

The pole and cart problem involves balancing a pole on a cart on a horizontal track that travels along a left-right axis. A human operator runs the pole and cart and is connected to a device that records the adjustments made to keep the pole balanced vertically. The human intervention forms a training set of data that contains rules that describe the adjustments.

Michie and Sammut applied the same process to controlling a satellite. They call these implementations "adaptive control systems." They point out that the conventional control theory requires a mathematical model to predict the behavior of a process so that appropriate control actions can be made. The relationships in the data were more complicated because, unlike the two-dimensional adjustments made to the pole and cart, a satellite has pitch, yaw, and roll directions. Nevertheless, the process is similar. A sample set of rules used to control a satellite's movements is shown in Figure 2.24:

Figure 2.24: Illustration of the Rules Used to Simulate Satellite Control

```
if ωz < -0.002 then apply torque of 1.5
else if ωz > 0.002 then apply torque of -1.5
else if pitch < -2 then apply torque of 1.5
else if pitch > 2 then apply torque of -1.5
else if ωy < -0.002 then apply torque of 1.5
else if ωy > 0.002 then apply torque of -1.5
else if roll < -2 then apply torque of 1.5
else if roll > 2 then apply torque of -1.5
else if ωx < -0.002 then apply torque of 0.5
else if ωx > 0.002 then apply torque of -0.5
else if yaw < -2 then apply torque of 0.5
else if yaw > 2 then apply torque of -0.5
```

Many processes are too complicated to model accurately. Often, not enough information is available about the process's environment. When the process is too complicated or the environment is not well-understood, an adaptive controller might work. An adaptive controller learns how to use the control actions available to meet the process's objective. The process is treated as a black box and the adaptive controller interacts with it by responses that have been learned through rule induction.

Rule Induction and the Work of Ross Quinlan

Rule induction was the inspiration for one of the most popular forms of machine learning, which was developed by Ross Quinlan at the University of Sydney, Australia. Quinlan developed one of the earliest top-down approaches to the rule induction of decision trees approach called ID3. ID stands for Interactive Dichotomizer and 3 stands for version 3, which was the most widely known version. More information and some useful references are provided by Michie (1991). Briefly, ID3 computes a gain ratio to determine the structure of the decision tree. The gain ratio functions like the variance reduction statistic in AID, or like the chi-square statistic in CHAID. ID3 is based on the concept of entropy, developed by Claude Shannon to describe the amount of information that is contained in a signal. Although this concept was originally used to describe the capacity of various communications channels, it can be used in decision trees to describe the communications capacity of competing splits or inputs and the resulting branches on the decision tree.

The ID3 algorithm had many of the same shortcomings as the AID algorithm; for example, decision trees might be grown too large to be reliable; multi-valued inputs could be favored over inputs with fewer values. Unlike AID or CHAID, ID3 did not combine similar values on the branches; if an input had three values, it produced a branch with three nodes, while a five-valued input produced five nodes, and so on.

The development of ID3 provided a significant boost for decision tree methods in machine learning. ID3 ultimately led to the introduction of more traditional, statistically based decision tree methods in a machine learning setting. This led to the continued development of decision trees in a variety of areas and applications—a development that continues to this day.

Improvements to the ID3 algorithm culminated in the development of the C4.5 method of decision tree construction, as well as its successor—C5.0. The C4.5 algorithm resolves problems identified in the original AID and ID3 implementations, and deals with both qualitative and quantitative attributes, missing values, and overfitting. C4.5 expanded the types of inputs possible—the target is nominal and the inputs can be either nominal or interval.

Unlike the ID3 algorithm, which produced *n*-way splits, the C4.5 decision tree algorithm produces binary splits. For multiple values, each attribute is first assigned to a unique branch, and then, in steps, two branches are merged until only two branches exist. Missing values are excluded from the split search on that input and from the numerator of the gain ratio. Missing values are an additional branch in the decision tree. For interval inputs, C4.5 finds the best binary split. For nominal inputs, a branch is created for every value, and then, optionally, the branches are merged until splitting does not improve the decision tree.

Merging is performed stepwise. At each step, the pair of branches that most improves the splitting measure is merged. When creating a split, observations with a missing value in the splitting variable are discarded when computing the reduction in entropy. The entropy of a split is then computed as if the split makes an additional branch exclusively for the missing values.

The decision tree is grown to overfit the training data. In each node, an upper-confidence limit of the number misclassified is estimated, assuming a binomial distribution around the number misclassified. A subtree is sought that minimizes the number of misclassifications in each node.

C4.5 can convert a decision tree into a *rule set*. A rule set is a collection of rules that describe the leaves of a decision tree. An optimizer runs through the rule set so that similar rules are combined and redundancies are eliminated. Because these rule sets contain fewer rules than the decision tree, they are more readily understandable than most rule representations. In some cases, rule sets can be more readily understandable than decision tree representations.

C4.5 can create fuzzy splits on interval inputs. The decision tree is constructed the same as with non-fuzzy splits. If an interval input has a value near the splitting value, then the observation is effectively replaced by two observations, each with weight related to the proximity of the internal input value to the splitting value. The posterior probabilities of the original observation equal the weighted sum of the posterior probabilities of the two new observations.

The most recent version of Quinlan's approach is C5.0. C5.0 is an improvement over C4.5 and provides boosting and cross-validation. Boosting resamples the data that is used to train the decision tree. Each time the data is used to grow a decision tree, the accuracy of the decision tree is computed. Over time, data is adjusted to address previously computed inaccuracies. C5.0 provides facilities to specify number misclassification costs.

The Use of Multiple Trees

The mid-1990s were a watershed era for decision trees. In addition to Quinlan's work and the work of Breiman, Friedman, Olshen, and Stone, significant developments came from computer science. Two researchers in particular developed a new approach that became influential. The work of Amit and Geman (1997) on digit recognition involved using multiple decision trees to create what they called a holographic view of the digits in the source database.

This led to the development of a new class of decision tree approaches based on resampling and reweighting the original data to create a family of predictors that perform together better than a single predictor. Multi-tree—or boost approaches—are discussed further in Chapter 4, "Business Intelligence and Decision Trees."

Amit and Geman compared their approach to creating a holographic view of the problem so that a given decision tree split could be viewed from various perspectives. The family of random decision trees that were grown from the original data is used to create these various perspectives.

Amit and Geman's work served as an inspiration to Quinlan, and most significantly to Leo Breiman, who developed this approach into random forests. The work on random forests is most notable because it represents decades of interaction between the machine learning community and the statistical community. This interaction is not always productive, possibly because of a difference in emphasis. As Breiman noted in the Wald series of lectures (2002): machine learning people tend to be interested in *whether things work*, whereas statisticians tend to be interested in *why things work*.

A Review of the Major Features of Decision Trees

So far, this chapter has described how decision trees have readily definable features that characterize and distinguish them from other data discovery, display, and deployment techniques. Decision trees were originally developed as robust yet simple methods to deal with the many complexities of multiple relationships among fields of information in data sets. These complexities and contextual effects are often missed by other methods of analysis, which can lead to inappropriate decisions. This is why decision trees, which explicitly discover and display multiple relationships in context, are such important tools for the empirical discovery, display, and validation of knowledge. The simplicity of decision trees facilitates the examination of multiple relationships, which enables decision trees to go beyond simplistic one-cause-one-effect types of analysis.

Roots and Trees

Decision tree results are produced graphically in the form of a decision tree. The normal display is an inverted tree with the root node at the top. The root node contains a summary of the data set to be examined; typically, it consists of the values of the field that will be partitioned or examined as the decision tree grows. Because this field is the target of the analysis, it is often called the target; however, because its values can be dependent on the values of the fields that will be used to examine it, then it can also be called a dependent field or variable.

Branches

Important inputs are selected as the splitting criteria in forming the shape and sequence of branches on the decision tree. The decision tree criteria separate important from unimportant branches so only strong relationships between inputs and the target are retained.

Inputs are referred to as predictors or classifiers because their values can be used to predict target values or classify target values. Whether inputs are predictors or classifiers, they are still considered inputs. Inputs have utility as a general descriptive term for predictors, classifiers, independent variables and, as is sometimes used in machine-learning applications, attributes. Branches can be 2-way (binary) or multi-way (many) and are formed by partitioning or splitting the target values with respect to the corresponding values in the inputs. Inputs can be any level of measurement—qualitative or quantitative.

Similarity Measures

Many measures have been used to select inputs and combine inputs that form partitions or classifications of the target. Attributes of the branches are grouped in the two or more nodes that characterize the branches. So, when a branch is identified with its associated leaves or nodes, then the members of each leaf or node are as homogenous as possible (with respect to their relationship with the target). In addition, each leaf or node is maximally distinguished or differentiated from other nodes on the same branch of the decision tree. Internode (between node) differences are maximized, and intranode (within node) similarities are maximized.

Typical statistical measures of association include a measure of how one set of values is related to or associated with another set; a measure of information gain (i.e., how much information about a target do I gain knowing corresponding information about an input?); or a measure of purity (how homogenous or diversified are the members of a branch of the tree?). It is possible to review the partitions or classifications formed by various inputs and to either select an input based on the numerical properties of the partitioning mechanism, or to select an input based on business rules.

Recursive Growth

Decision trees are said to be grown recursively; that is, once the initial or root node is split into a branch, all subsequent nodes are also split using the same methodology. So, once you discover how to split one node on a decision tree, you can recursively apply the same methodology to all descendent branches and associated nodes on the decision tree. Once a classification that is formed by the branches of an input is selected, the decision tree can be grown incrementally by descending to the nodes formed by this branch. This branch is, in turn, partitioned like the original root node was. This process continues as the decision tree is grown until it either runs out of data in the descendent node, or the growth is stopped according to a stopping rule. This is called recursive partitioning growth.

Shaping the Decision Tree

Various stopping rules can suggest when recursive partitioning should be stopped. It is necessary to stop at some point because deep decision trees are more complicated to understand and less useful. In addition, the lower branches are formed by fewer cases in the target data set; therefore, the statistical results are based on less statistical power and are consequently less accurate and reproducible. The validity, accuracy, and reproducibility of the decision tree can be tested through validation. Indeed, both validation testing can be used to shape the form and depth of the tree (including which input to use for branching and how many branches to form for a given input).

Deploying Decision Trees

The results, interpretation, and application of decision trees can be described semantically as simple IF <condition> THEN <action> rules. This way of describing relationships is very general and close to natural language, so it is readily understandable in non-scientific (i.e., non-mathematical) situations. In fact, these rules are virtually indistinguishable from the programming rules in many programming languages. In most cases, the rules are deployed in a markup language, such as PMML (Predictive Modeling Markup Language).

A Brief Review of the SAS Enterprise Miner ARBORETUM Procedure

The SAS ARBORETUM procedure is the computational engine that lies behind the decision tree construction that is found in SAS Enterprise Miner. The ARBORETUM procedure works with nominal, ordinal, and interval data as both inputs and targets in a decision tree. The ARBORETUM procedure forms branches in a decision tree using a variety of criteria, including:

- variance reduction for interval targets
- F-test for interval targets
- Gini or entropy reduction (information gain) for categorical targets
- chi-squared for nominal targets

The ARBORETUM procedure produces both binary and multi-way branches in the decision tree. Missing values in the input fields that are used to form branches can be handled in a variety of ways:

- use missing values as a separate, but legitimate code in the split search
- assign missing values to the leaf that they most closely resemble
- distribute missing observations across all branches
- use surrogate, non-missing inputs to impute the distribution of missing values in the branch

The ARBORETUM procedure provides a variety of methods for trimming and shaping the size and form of the decision tree, including:

- cost-complexity pruning and reduced-error pruning (with validation data)
- prior probabilities can be used in training or assessment
- misclassification costs can be used to influence decisions and branch construction
- interactive training mode can be used to produce branches and prune branches

The ARBORETUM procedure provides methods to compute variable importance, which can be done with both training and validation data. The ARBORETUM procedure provides for the generation of SAS programming code. This code can contain indicator variables that refer to specific leaves on the decision tree. These indicator variables can then be used as inputs to capture effects in other modeling applications. In addition, the generation of PMML code is provided.

---

[1] Figures presented in this example are, in general, the same as those in the original article. The variable names and scenario have been changed to reflect a marketing application instead of the epidemiological research application that was featured in the original article.

[2] Michael Weisberg, John Krosnick, and Bruce Bowen provide a more recent description of this method in *An Introduction to Survey Research and Data Analysis* (1989); however, the basic methodology remains unchanged to this day.

[3] When a decision tree fits a linear relationship, it tends to fit the single line—represented by a slope coefficient in regression—as a series of decision tree branches. This tends to produce a line-fitting, staircase effect, which is neither economical nor as effective in prediction as regression is. Recent developments in multi-tree techniques, discussed elsewhere in this book, offset this disadvantage somewhat.

Chapter 3: The Mechanics of Decision Tree Construction

The Basics of Decision Trees

The goal of this chapter is to provide a comprehensive and detailed overview of the process of growing a decision tree. Many of the most common decision tree options and approaches are covered. These options and approaches have their roots in the original AID algorithm, as well as successor algorithms, such as CHAID, ID3, and CRT. The decision tree component of SAS Enterprise Miner incorporates and extends these options and approaches. It includes the popular features of CHAID and CRT and incorporates the decision tree algorithm refinements of the machine learning community (including the methods developed by Quinlan in ID3 and its successors: C4.5 and C5).

The SAS Enterprise Miner decision tree supports both interactive (manual) and automatic growth approaches. Adjustable defaults are provided in both interactive and automatic approaches to help identify the best decision tree models for the analyst's purpose.

The decision tree growing process can be broken down into a number of subprocesses, as shown in Figure 3.1.

Figure 3.1: Illustration of Subprocesses in Growing a Decision Tree

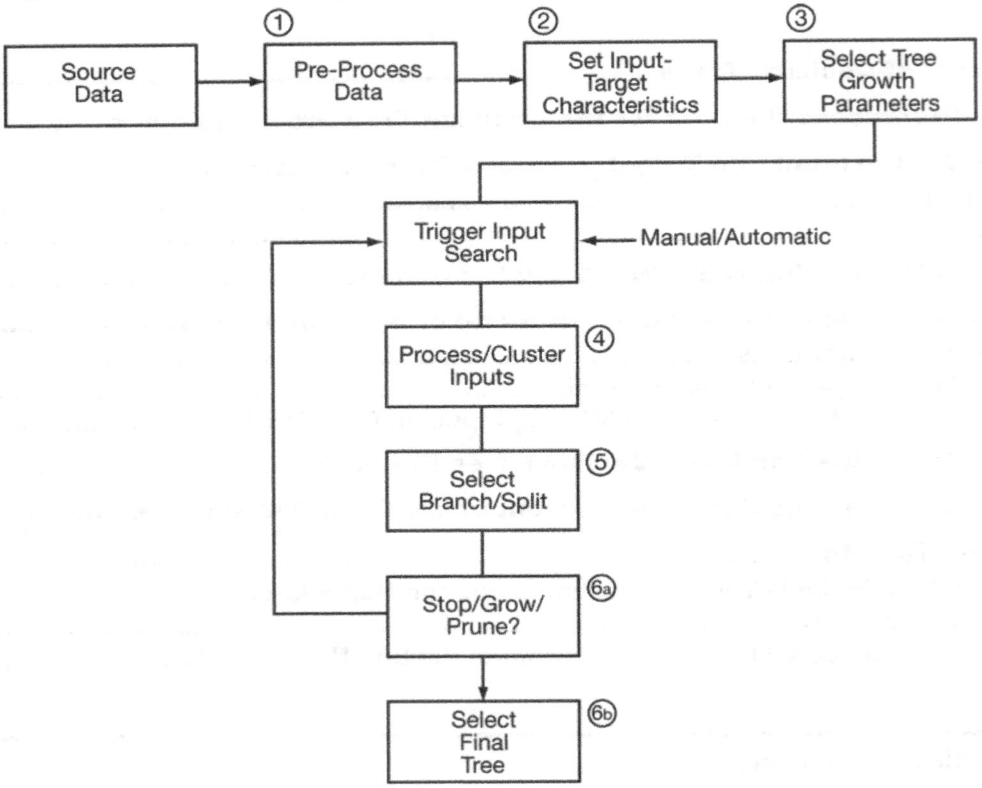

The six steps for growing decision trees are:

1. Preprocess the data for the decision tree growing engine.
2. Set the input and target modeling characteristics.
3. Select the decision tree growth parameters.
4. Cluster and process each branch-forming input field.
5. Select the candidate decision tree branches.
6. Complete the form and content of the final decision tree.
   a. Stop, grow, prune, or iterate the decision tree.
   b. Select the final decision tree.

These steps are performed in sequence, with the development of each layer of branches (or levels) of the decision tree. The decision tree growing process—steps 4 and 5—is an iterative process. This means that once the steps have been applied to the main set of data, which forms the root node of the decision tree, they can be reapplied recursively to any descendents of the root node.

Step 6—Complete the form and content of the final decision tree—is subject to both formal and informal shaping methods, which are used to terminate tree construction often before the mechanical components of the tree-growing algorithms stop functioning.

Step 1—Preprocess the Data for the Decision Tree Growing Engine

Data preparation is a study in its own right. There are books and courses on data preparation [for example, *Data Preparation for Data Mining* (Pyle 1999)]. It is frequently necessary to write code to preprocess the data. For example, the following SAS code transforms string abbreviations into numeric state codes:

```
IF substr(upcase(left(state)),1,2) in ("ME","NH","VT","MA")
THEN region = 1
```

Here are some rules-of-thumb for decision tree modeling.

- Understand the differences between categorical and continuous data. Categorical data such as ZIP codes might have a numeric form with many values that can look like continuous data, but that are actually categories. Consider clustering categories together outside of the decision tree. It might be possible to cluster categories together with respect to a target variable (this is discussed later).
- Categorical targets with more than two values are extremely difficult to interpret. Rework multi-category targets into a 1-of-N code scheme. With 1-of-N coding, each distinct category becomes a binary yes-no/on-off outcome in a new input. There are as many binary inputs as categories in the original multi-category input.
- Dates can be a continuous field, but might need to be changed to Julian format. It is useful to compute time intervals, such as length of time as a customer, before beginning analysis.
- There can be other time and distance measures; these need to be calculated and verified before analysis.
- Try to avoid information loss; higher levels of measurement contain more information than lower levels, so actual income is preferable to income ranges.
- If you are working with variables that are expressed along a scale (for example, 1, 2, 3, and so on), then you might find it easier to express all scales in the same direction.
- Multiple response items might need to be treated with care. For example, if you have a list of products that are purchased, then each product might need to be totaled separately, and a total number of products purchased might need to be calculated. In this situation, multiple response items within each unit of observation might need to be summed to create an analysis data set.

- Do not confuse missing information with a missing value because this is not always the same. For example, Quantity Purchased can be blank for a given day or a given product type if the customer did not purchase on that day or did not purchase that product.
- It might be necessary to pivot records, particularly if you want to compute purchase quantities for given products. The product purchases tend to be one-line-per-purchase records with purchase details and a customer number. The purchase details need to be summarized through a pivoting operation, such as PROC TRANSFORM. Then, the aggregates are attached to the record (typically using customer number as a key).

Once the data is available, display the attributes using a summarization routine, such as what is provided in the StatExplore node in SAS Enterprise Miner. The StatExplore node produces a good diagnostic summary of the attributes, as illustrated in the following output using the shopping data set from Chapter 2.

Variable	Numcat	NMiss	Mode	Pct	Mode2	Mode2Pct
RECENCY	8	0	30	20.17	60	17.38
children_home	3	10822	64.43	Y	33.17	
freq	7	0	5	47.59	4	23.21
gender	3	1839	female	58.95	male	30.1
has_new_car	2	0	N	63.51	Y	36.49
inc	10	2622	$100,000- $124,999	22.97	$75,000- $99,999	19.91
maritalStatus	2	0	Married	63.31	Self	36.69
money	18	0	200-299	16.12	100-149	15.19
occupation	10	7459		44.41	1	22.74
state	46	0	NJ	11.33	CA	10.31

Class Variables 1

Variable	Mean	StdDev	Non Missing	Missing	Min	Median	Max
NetSalesLife	248	283	16797	0	-4991	168	
8493							
adultsInHH	2	1	16797	0	0	2	6
age	47	13	15693	1104	18	46	96
bathroomPurchases	5	8	16797	0	0	2	240
bedroom	3	4	16797	0	0	1	74
couponPurchase	2	5	16797	0	0	0	105
display	2	3	16797	0	0	0	78
hasBankCard	1	0	16797	0	0	1	1
hasStoreCard	1	0	16797	0	0	1	1
has_card	0	0	16797	0	0	0	1
has_credit_card	0	0	16797	0	0	0	1
has_upscale_store_card	1	0	16797	0	0	1	1
income	6	2	16797	0	0	7	9
juvenile	2	4	16797	0	0	0	129

kitchen	2	4	16797	0	0	1	96
length_of_residence	7	5	15356	1441	1	6	15
lifeTransactions	21	18	16797	0	3	16	379
lifeVisits	4	2	16797	0	3	4	50
mystery_field	1	0	16797	0	0	1	1
owns_RV	0	0	16797	0	0	0	1
owns_motorcycle	0	0	16797	0	0	0	1
owns_truck	0	0	16797	0	0	0	1
table	1	2	16797	0	0	0	37
topIndicator	1	0	16797	0	0	1	1
valueOfCar	24	19	7743	9054	1	20	205
windowDisplay	5	6	16797	0	0	3	85

Interval Variables 1

When defined and introduced into the data mining environment, the data set takes the form of a table with rows and columns. The rows represent individual records or observations. The columns contain measurements taken across each record or observation. So, each data line represents an object of analysis that has attributes with associated values.

Step 2—Set the Input and Target Modeling Characteristics

Decision tree inputs and targets can be encoded at any level of measurement, ranging from raw, nonmetric categories (such as high, medium, and low) to highly refined, precise, quantitative measurements (such as temperature in fractional degrees of Fahrenheit or Celsius). It is useful and necessary to preprocess the inputs and targets in order to do meaningful work with a decision tree. This step is typically a refinement of the tasks described in Step 1 above and is equivalent to fine-tuning the representation of the input fields to fully support the partitioning algorithms that are run by the decision tree node. As you will see below, many of the steps are carried out automatically by the software.

One of the fields of the data set serves as the target of analysis. Other fields are defined as inputs that can be used to predict or describe this target of analysis. These inputs are columns of the table that are used as input variables to construct a set of decision rules. These decision rules describe or predict variations in the target.

Targets

Interval targets are the easiest to deal with. Almost all decision tree algorithms accept an interval target. In a data analysis task, you should always check the missing value indicator for the target field. Look for values such as –1, –99, and even 0 (which can indicate a missing measurement) and ensure that the target field is either removed from the analysis or handled appropriately.

Some fields such as SIC code or ZIP code can appear to be interval targets, when in fact they are categorical targets and, except under special circumstances, cannot be treated as interval targets. Variables can be treated as interval targets only if the average value and deviations from the average value have meaning. In the case of SIC codes and ZIP codes, the average value of SIC codes or ZIP codes does not represent the average value of the codes that they are formed from. Therefore, averages and deviations from averages have no readily interpretable meaning for these codes.

With this type of categorical target, you create a 1-of-N derivation of the categorical codes. Thus, SIC code 8062 (hospitals) becomes 1 and all other target values become 0. This enables you to distinguish hospitals from all other target values in the analysis. Similarly, ZIP code 10010 (for New York city) becomes 1 and all other target values become 0.

You should treat all categorical outcomes this way, even when a small number of categories could be modeled as an unmodified categorical or nominal target, such as a target with low, medium, and high category codes. (In this case, the target is also an ordered target.) The main reason for this recommendation is interpretability; it is very hard to understand the distribution of categories in the nodes of the branches of a decision tree when more than two categories are present.

The decision tree algorithm accommodates multi-valued categorical and ordinal targets so you can use them in the raw form. If you are interested in prediction, then the ability to read and interpret the nodes of a decision tree might be less pressing. Consequently, in predictive applications, the need to change multi-valued categorical and ordinal targets might not be great.

In the following example, height is the target variable that is set up as a function of various input variables (or inputs), which are used as distinguishing attributes to construct the decision rules that describe the functional relationship between target and inputs.

```
Height <--- I1, I2, ... , In  (For example, Gender, Age, ... , Hair
color)
```

Height, which is the target, is called a target or dependent variable (meaning, because its value is a function of the input variables, it depends on the values of the inputs). Because the values of the inputs in this formulation can vary, they can be called independent variables.

Targets can be quantitative or continuously valued entities, such as height. Targets in nonmetric, nonquantitative, or categorical forms (such as short or tall) can also be used. As with targets, input fields can be any measurement, from nonnumeric categories (such as gender) to numeric quantities (such as weight and age). Numeric quantities can be used in both continuous and ordinal representations. For example, you can have age ranging from 0 to infinity (quantitative representation) or in one-year increments, such as 1 through life expectancy < 100 years.

Both targets and inputs can be unordered, such as the State code. Decision trees are a useful and versatile method for handling many unordered inputs. Although decision trees have many flexible and powerful ways of handling multi-category targets, decision trees with more than two or three categories become difficult to work with and interpret.

Inputs

In the terminology that we use here, fields used to define the branches of the trees are called "inputs". Like all the other fields discussed here, these inputs can have different levels of measurement:

- binary
- class levels or categories
- ordinal
- interval

The branches on a decision tree are said to be formed when the values of a node (beginning with the root node) are split into branches. The splits are formed by partitioning the range of values in the input that is selected to split the node.

Recall that inputs can form branches may be binary or multiway. Some further thoughts on using inputs to build decision trees are presented below:

Pre-Process the Data

The decision tree algorithms are exceptionally robust in dealing with various kinds of input data. In particular, the decision tree algorithms include mechanisms to deal with missing values. Settings for input level of measurement, as shown in Figure 3.2, are made in the "Edit Variables" selection for the SAS Enterprise Miner table properties.

Figure 3.2: Adjusting Level of Measurement in the Data Source

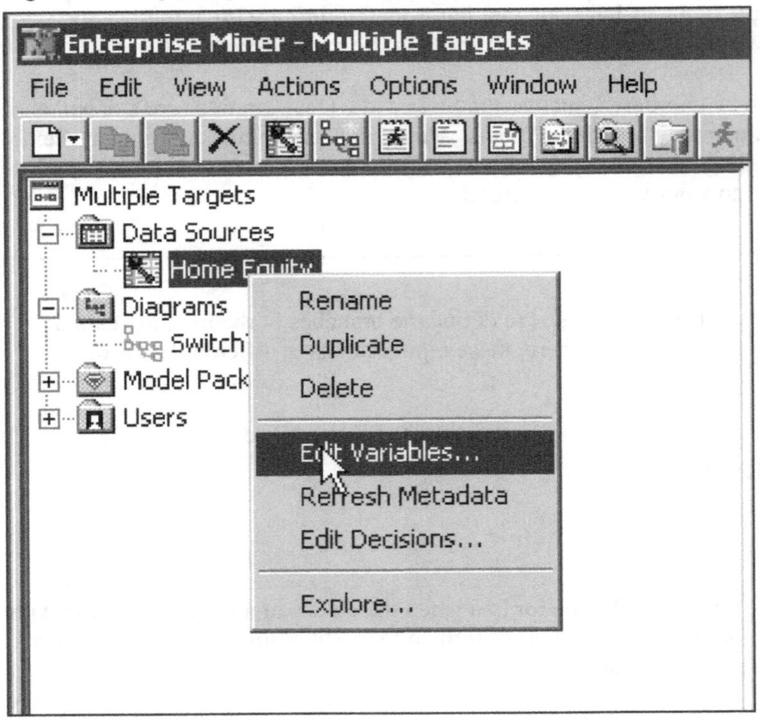

Usually, a binary input has a binary setting. An interval input has an interval setting. Class inputs usually have a "Nominal" setting. This indicates that values combine in a variety of ways in order to form the branches of a tree. Occasionally a researcher might decide to include a class input that has a restricted number of values as an "interval" input. In this case, the values are determined to be monotonically increasing.

Automatic Computation of Input Split Categories

Generally, for a binary split, the interval inputs are never discretized. For a multiway split, X is discretized into 100 categories by default (option INTERVALBINS=). To define the categories, the decision tree algorithm first finds the worth of every possible binary split. Then it keeps the best 99 of these that are not too close to each other. This discretization is done separately in every node.

Interval inputs can be enumerated computationally to find the best split. The usual enumeration method is to calculate the log-worth statistic and to find groups along the range of interval values that produce the best split worth. When binary splits are selected, then only one cutpoint on the range is selected so as to form two descendent branches. When multiple branches are selected, two or more cutpoints are calculated and the highest-worth partitioning is used to display the results.

To speed computation, a sample is taken each time the inputs to a node are examined for potential splits. The sample size must be large enough to produce reliable results. A sample size of just over 30,000 is used each time a new node is split (and new samples are drawn so that sampling bias is randomized throughout the final decision tree).

In the case of ordinal or class inputs, there are a large number of potential partitions that could be identified to form branches. This can be quickly seen in the case of class inputs where discrete values can be formed in any order. Many more combinations are possible than with interval inputs (where only ordered combinations of values are allowed). By one calculation, there are 2^{n-1} combinations of partitions for class variables, so a class variable with 10 values has 512 possible combinations. This number of combinations increases rapidly with the absolute number of discrete values. The large number of potential groupings is identified by means of a heuristic that first consolidates large numbers of discrete categories into "consolidation groups" that are identified as likely to share the same value for the target field. This heuristic reduces the number of potential combinations that must be searched in order to find high log-worth branches.

Preempting Automatic Treatment of Input Values to Split Categories

Sometimes a researcher wants to preserve or present a particular range of values for the numeric fields that are included in the analysis. In this case, we might be interested in directly assigning values to ranges ourselves. We might want to rework numeric fields as follows:

- in defined ranges (as in the example of collapsing age into the range 15–25)
- as buckets or bins, in which the number of records in each bucket or bin depends on the definition of the bin attributes and on the mapping process that assigns individual records to each bin in the data set
- as deciles or other quantiles, in which there are an equal number of records in each decile group

The Modify category of SAS Enterprise Miner tasks provides a number of data modification nodes that perform the kinds of pre-processing transformations that might be required. For example, the SAS Enterprise Miner Transform Variables node can undertake a variety of transformations. (See the latest SAS Enterprise Miner documentation at http://support.sas.com for details). For example, this node can take a parameter that determines the number of bins (called n in the following equation). This parameter indirectly specifies the minimum width between two successive candidate split points on an interval input. The width equals $(\max(x) - \min(x))/(n + 1)$, where max(x) and min(x) are the maximum and minimum of the input variable values in the within-node sample being searched. The width is computed separately for each input and node.

You could classify a field like age into 10 buckets or bins to turn it into a categorical field. It is also possible to define your own preset collapsed categories. For example, age can be classified into the following groupings:

Preteens	< 13
Teens	13–19
Young adults	20–29
Older adults	30–55
Seniors	> 55

If a quantile has been defined, then binning is based on the frequency of records that are in a quantile range. Quantiles are computed by taking the frequency of each value in the input. The quantile is formed by establishing a lower and upper quantile value that encompass the number of records that form the quantile. So, if a decile is computed, then the first decile contains 10% of the input records according to the frequency of occurrence.

In addition to establishing all input values as categorical fields, you must establish whether the categories are ordered or unordered. Age is ordered or monotonic—there is a steady increase in the magnitude of age as you move from the lowest category (preteens) to the highest category (seniors).

However, if you use the values of State as input, then you would define this input as nominal or unordered. There is no innate underlying increase or decrease in the assumed magnitude of a State code as you move from IA to AL, or, if the State codes are assigned a number, as you move from 12 (representing the state of New York) to 25 (representing the state of Montana).

Step 3—Select the Decision Tree Growth Parameters

Although there are many decision tree algorithms, forming a decision tree is a simple process. Originally, decision trees were formed by sorting inputs into ordered groupings. The sort order was a function of the value of the input category, with respect to the value in the target. With a continuous target, input categories were sorted from low to high, and the corresponding average target value was recorded in each of the input categories. If the input categories were combined to form a node that represented multiple categorical values, then the input categories were combined either ordered or unordered.

This original decision tree process begins by transforming each ordinal and interval type of input into categories that can be manipulated and combined.

Once all inputs are transformed, the decision tree algorithm performs its most important task—picking the best input to form a split. A number of choices affect this task. The more important considerations are the following:

- How will input categories be combined to form branches, or will they be combined at all?
- How will branches be sorted and combined? Will they be in line with their level of measurement—continuous, ordinal, and categorical?
- How many nodes on a branch will be allowed?
- How many alternative branches will be stored at each level of the decision tree?
- How will differences be determined (for example, predictive power between branches)?
- How will branches be evaluated, selected, and displayed?
- How will input data records be segmented into branches? What will happen when a given input data record has missing information in the input field that is being used to form a branch?
- Will a branch growth strategy be based on empirical tests of accuracy or will theoretical tests (for example, hypothesis tests) be used to grow branches?
- Will branches be pruned ahead of time or will branches be trimmed after an initial growth stage or once an overly large tree has been grown?
- When will the decision tree processor stop identifying potential branches?
- When will the decision tree stop identifying potential nodes?

A number of settings determine how to act on these considerations. For example, the total number of potential split-and-merge points in a range of bins for an input is an important setting. A split-and-merge point is a potential cutting point between two bins or categories of an input that can be merged to form a larger bin. Because the bins have different target attributes (which means they have different input-target behavior), categories are split apart. A typical default is to examine 5,000 potential split-and-merge points. If there are 5,000 or fewer points to examine, then SAS Enterprise Miner exhaustively examines all possible points. If more than 5,000 points exist, then a heuristic merge-and-shuffle approach is used. A merge-and-shuffle approach examines approximate groupings of categories so that the branches capture differences in the input-target behavior.

Various measures of computing the strength of a branch are used. Branches can be picked based on strength or based on validation and test statistics. The strength value is referred to as the "worth" of a branch. Typically, higher worth branches are picked over lower-worth branches.

The decision tree mechanism in SAS Enterprise Miner treats missing values in different ways. The mechanism can: 1) include missing values, 2) put missing values into a separate category, 3) distribute observations with missing values in proportion to the size of the nodes on a branch, or, 4) use surrogate inputs in place of missing values. A surrogate input performs like an input, but has a lower worth. A surrogate input is highly correlated with the missing value, and forms the branch partition when there is a missing value for the input on the data record that is making the split. In this case, a non-missing surrogate input value on this data record can be used. You can set the maximum number of surrogates that can be used in growing the decision tree because when surrogates are used, the data has to be re-read (and this takes more time).

Nodes can be constrained to a minimum size, and any category in the target classification can be constrained to a certain size. For example, a constraint can be that no nodes with less than 50 observations will be identified, and no node with a categorical value with less than five observations will be identified.

Decision trees can be constrained to grow to a certain depth. For example, decision trees can be constrained to stop growing after three levels. Levels are calculated from the root node (in the example, the first set of branches forms the first level).

Inputs can be constrained to be used on only one level, so once they are used, they cannot be used again on a lower level.

Step 4—Cluster and Process Each Branch-Forming Input Field

Clusters of codes are formed when values in the input fields that form the branches of the decision tree are grouped together. The goal of this step is twofold:

1. Put similar observations into the same cluster so that the characteristics of the observations in that cluster are as similar as possible. Create clusters on the same level of the split on the tree so that the differences between clusters are magnified. Create clusters of codes that make intuitive or theoretical sense (as when State codes are arranged into East, Central, West, North, and South).
2. Clustering forms nodes on the input that maximize the predictive relationship between the input and the target. This is the original and persistent goal of forming leaves on the input. However, this goal is sometimes at odds with the first goal because the most appealing or most understandable branch is not always the best predictor. In this case, the analyst has to choose between numerical strength and interpretability.

The result of the clustering step is the selection of an algorithm and the guidance rules that form the branches that, in turn, form the decision tree. One exception to this step is the rare case in which no input codes are grouped together (such as when the ID3 algorithm is running). The setting in SAS Enterprise Miner that controls this step is the maximum number of branches. For example, a setting of 2 to 50 will accommodate the construction of branches with 2 to 50 nodes or leaves.

When the maximum number of branches is set to 2, the decision tree is a binary tree. A number greater than 2 results in multi-way branches. Binary branches are easier to calculate. Input values are clustered on one of two sides when forming the decision tree branches—either on the left side or on the right side. The main question is whether the categories of the inputs are clustered in an ordered or unordered way. With ordered comparisons, adjacent categories in the range of categories in an input can be combined. With unordered comparisons, any category in an input can be combined with any other category. These different comparisons produce different results, as shown in Figure 3.3 and Figure 3.4.

In Figure 3.3, a decision tree shows the relationship with YOJ (Years On the Job) and its influence on the target variable in the analysis—bad debt. This decision tree is from the Home Equity Loan data that is provided in the sample Help files with SAS.

Figure 3.3: Illustration of Clusters Producing Two Nodes

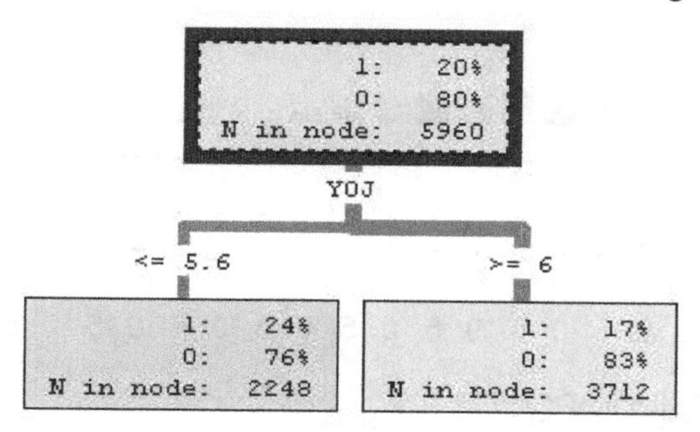

All values lower than or equal to 5.6 are on the left side of the branch; all values greater than or equal to 6 are on the right side. In Figure 3.4, an unordered search for branches in the YOJ variable is shown.

Figure 3.4: Illustration of Binary Partitioning with Unordered Branch (Cluster) Components

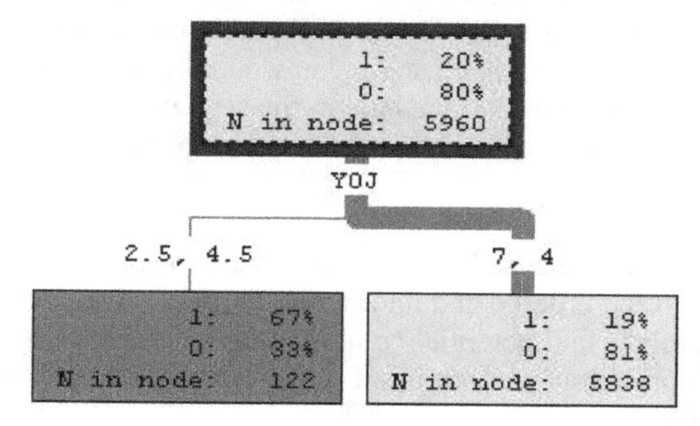

When unordered clustering is used, out-of-sequence values can combine. In Figure 3.4, the value of 4, which would usually be in the sequence between 2.5 and 4.5 on the left side of the decision tree, has been grouped with the value of 7 on the right side of the decision tree. If you look at the distribution of YOJ versus bad debt, as shown in Figure 3.5, you can see this process at work.

Figure 3.5: Illustration of Bad Debt Distributed by YOJ - Ordered Search

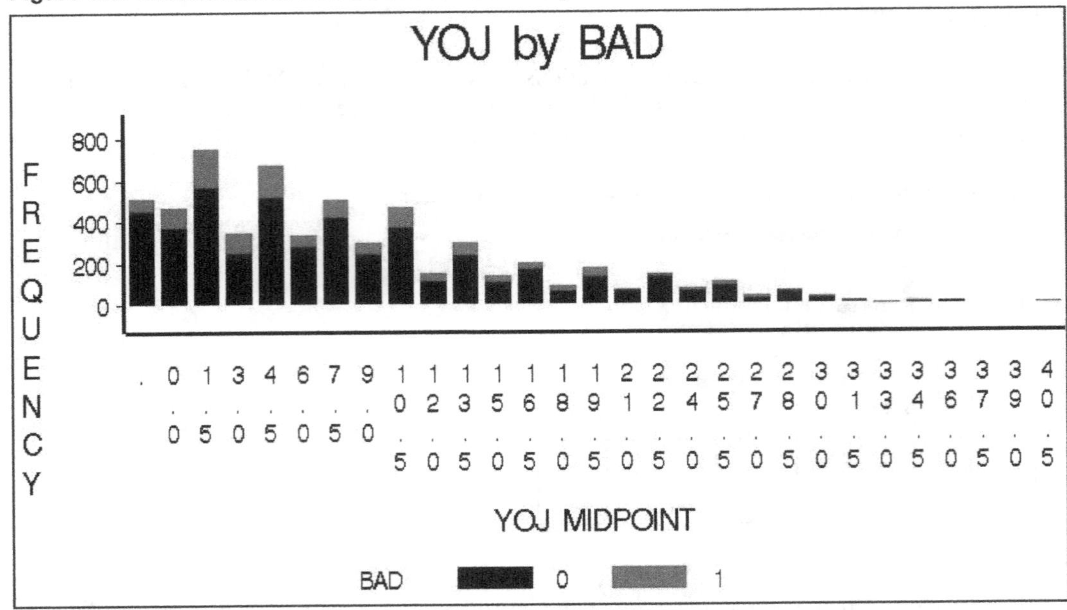

The first five vertical bars have an elevated BAD count when compared to the bars that follow (in this example, BAD=code 1). In an ordered comparison, you would expect to see a split in the left and right branches, between the fifth and sixth bars. At this point, the cumulative number of bad counts is greater than the cumulative number of bad counts to the right (i.e., the sixth bar and greater). This is exactly what is in the decision tree with the ordered split in Figure 3.4 (the split is at value 5.6). However, in an unordered search for left and right clusters, you would see that the fourth bar has a lower proportion of bad debts when compared to the bars in the sequence of low values less than 6. The unordered search algorithm has determined this and has grouped the value 4 on the right side of the decision tree.

All of these results are preliminary and unverified and are based on samples, not on the data set. They are provided to illustrate the differences between ordered and unordered searches. After you begin to examine the distribution of values, you might find potential combinations of values that SAS Enterprise Miner has not detected. The comprehensive search of all possible values is covered in the discussion of search algorithms that follows.

Clusters of input values that form multi-way branches are more difficult to calculate than binary branches. However, multi-way branches have useful visual and conceptually appealing properties that make them more robust, economical, and flexible, especially in explanatory and expository decision tree applications. To explain further, suppose a decision tree breaks out sales into West, South West, Great Lakes, North East, and South East regions. All regions can be represented on one level of the decision tree with one decision rule. Binary trees would need to segment the data four times, with corresponding decision rules, to accomplish this. With multi-way branches, however, there are the issues of how many multi-way nodes to create, and where to establish the cutting points in the clusters that form the multi-way nodes.

Clustering Algorithms

A variety of clustering algorithms can form the branches that define the leaves in the decision tree. All algorithms try to create leaves that are as alike as possible (i.e., pure or homogeneous) and that are as different as possible when compared to other leaves on the same level of the decision tree. Observations in a leaf are similar, and differences between observations in leaf-to-leaf comparisons are as great as possible. Choosing an algorithm depends on the measurement of the input values that are grouped together:

- continuous-type measurements need quantity measures of clustering (to determine similar group members)
- categorical-type measurements need categorical measures of clustering (to determine similar group members)

The clustering algorithms most commonly used are the following:

- variance reduction (used in the original AID)
- entropy (adopted by Ross Quinlan)
- Gini (introduced by Breiman et al.)
- tests of significance (introduced by Kass)
    - t-test and F-test for continuous measures
    - chi-squared test (Fisher's exact test for small numbers) for categorical measures

When tests of significance are used, Bonferroni adjustments can be applied. Bonferroni adjustments compensate for measurement and test effects that force changes in the one-test/one-hypothesis approach. This approach forms the basis of traditional tests of significance and, consequently, the statistical tables that are published that reflect the underlying mathematics of these tests. This multi-test adjustment was originated by Kass (1975). It addressed shortcomings with the original AID algorithm and with the variance reduction test that was used to form the resulting decision tree. Specifically, these tests do the following:

- compensate for multiple tests of significance (affect the presentation order of potential branches)
- adjust the statistical strength of any input in the analysis to compensate for the number of inputs that are used to form potential branches (for example, asking "Is it a significant relationship?")
- adjust the stopping rule that is used to stop growing the tree (if a statistical test of significance is used as a stopping rule)

Variance Reduction

After the data is preprocessed (so that, for example, all continuous inputs are arranged as categorical inputs), the effect of each input is examined. The goal is to cluster the attributes of the inputs together, based on the strengths of their relationships with the values of the target. In the original AID decision tree approach, variance reduction was used to form binary groups or clusters for each input in the data set. Values were chosen so that variations from the average values of the two groups formed by the binary branch were minimal.

Gender	Height	Weight	BMI
male	68	203	31
female	59	94	19
female	64	113	19
male	64	160	27
female	67	125	20
female	64	120	21
female	64	120	21
female	67	134	21
female	63	125	22
male	65	135	22
female	65	135	22
male	67	144	23
female	67	145	23
female	57	105	23
male	68	150	23
female	68	150	23
male	66	143	23
female	62	128	23
male	71	170	24
female	63	138	24
male	65	148	25
female	62	135	25
male	65	155	26
male	65	160	27
female	62	97	18
male	67	175	27
male	63	160	28
male	62	155	28
male	64	180	31

In the previous display, the mean and variance of BMI (Body Mass Index) are 23 and 11, respectively. BMI is calculated as weight divided by height, squared. Variance is the sum of the deviations of the individual measurements around the mean or average of the measure. If BMI is segmented into males and females, then the mean and variance are as follows:

	Number	Variance
Female	21	4
Male	26	9

Segmenting the BMI scores into females and males can reduce the variance in the resulting groups when compared to the overall variance. Gender is an important discriminator and is a likely candidate for splitting criterion in the decision tree.

Gini Impurity

The Gini index is a measure of variability/purity for categorical data. It was developed by the Italian statistician Corrado Gini in 1912. The Gini index can be used as a measure of node impurity, where $p_1, p_2, \ldots, p_r$ are the relative frequencies of each class in a node. The Gini criterion was proposed by Breiman et al.

$$\textbf{Gini impurity} = 1 - \sum_{j}^{r} p_j^2$$

A pure node has a Gini index of 0—as the number of evenly distributed classes increases, the Gini index approaches 1. The splitting criterion is the one that most reduces the node impurity. In the following example, the impurity of the root node when considering body type is shown:

1 – (average purity measure) – (heavy purity measure) – (slim purity measure)

The Gini index is computed as follows:

$$1-(8/28)^2-(11/28)^2-(9/28)^2 \quad \text{or } .82$$

Gender	Weight	Height	Ht_Cent.	BodyType
Female	89	5'3	160	slim
Female	117	5'7	170	slim
Female	128	5'2	157	slim
Male	132	5'1	155	slim
Female	150	5'2	157	slim
Male	150	5'2	157	slim
Female	160	5' 4	163	slim
Female	179	4'10	147	slim
Female	167	5'3	160	slim
Male	161	5'6	168	average
Male	163	5'7	170	average
Male	180	5'4	163	average
Female	167	5'1	180	average
Male	188	5'6	168	average
Male	191	5'8	173	average
Male	194	5'7	170	average
Male	206	5'4	163	average
Female	215	5'2	157	heavy

Male	201	5′7	170	heavy
Female	182	6′2	188	heavy
Male	201	5′9	175	heavy
Male	206	5′9	175	heavy
Male	206	6′	183	heavy
Male	216	5′9	175	heavy
Male	239	5′5	165	heavy
Male	220	6′1	185	heavy
Male	254	5′8	173	heavy
Male	284	5′6	168	heavy

A split on gender produces two nodes with impurity measures of .48 and .89 (for females and males, respectively). This is a reduction in impurity of .41.

Entropy and Information Gain

Entropy was developed as a measure of the uncertainty of a transmitted message, in bits (Shannon and Weaver, 1949). Entropy is used with categorical outcomes—it measures variability (homogeneity) in splits and the leaves that are formed by the splits.

The entropy of a split is found by computing the entropy in each of the leaves, and by summing the entropy of all the leaves of a split. The variability of an outcome in a leaf is computed using the formula **$-\log^2(p_i)$**. The summed entropy of all the leaves of a split is

$-\sum_{i=1}^{r} p_i \log_2(p_i)$, where $\mathbf{p}_i$ is the proportion of a particular class ***i*** in the collection of categories contained in the branch.

This measure is calculated as follows. The decision tree has 28 observations. The probability of body type is 8/28 (0.29), 11/28 (0.39), and 9/28 (0.32), for average, heavy, and slim, respectively.

To compute the entropy, sum the three $\sum_{i=1}^{r} p_i \,\mathbf{\log_2(p_i)}$ terms. The results are –0.51639, –0.52954, and –0.52632, with a summed entropy of 1.57. You can use other splits among the categories in the table, such as gender. For example, the entropy for males is 1.382 and for females is 1.156. The total entropy for gender is 1.30 and the information gain is .27 or 17%. You can verify these results with the following display:

Body Type	Frequency	Percent	Cumulative Percent
average	8	28.57	28.57
heavy	11	39.29	67.86
slim	9	32.14	100
	28		

For body type, the best guess for the 28 observations is **heavy**. The likely outcome to classify the body type is 11 right guesses versus 17 wrong guesses, which results in a classification hit rate of about 40%.

bodytype	Gender(Gender) Female	Male	Total
average	1	7	8
	3.57	25	28.57
	12.5	87.5	
	10	38.89	
heavy	2	9	11
	7.14	32.14	39.29
	18.18	81.82	
	20	50	
slim	7	2	9
	25	7.14	32.14
	77.78	22.22	
	70	11.11	
Total	10	18	28
	35.71	64.29	100

Knowing the distribution of gender increases the ability to guess correctly. For males, the best guess is **heavy**; for females, the best guess is slim. Using this guessing strategy means that you could get 9 guesses for males wrong, and 3 guesses for females wrong, which results in a classification hit rate of 57%. This yields an improvement of about 40%, equal to the information gain previously calculated.

Chi-Squared

The clustering process in the CHAID approach to forming a decision tree is based on applying a test of significance. In CHAID, input groupings are formed by combining values in the input if their relationships with the target are similar. Values are indistinguishable from a statistical point of view if the pairwise differences between two values relative to the target are not statistically significant. Statistical significance can be determined by a simple t-test that tests the differences between the average values of the target for one input value versus another input value. If the test is not significant, then the two values are combined.

Ordered comparisons require pairwise comparisons of adjacent values. If the test of significance fails, then the values are combined to form one category.

Unordered comparisons require pairwise comparisons of all available values, regardless of order. If the test of significance between the two selected values fails, then the values are combined to form one value.

When selecting the test of significance, the test that is applied depends on how the values of the target are measured. Are they categorical (for example, yes–no) or are they numeric (for example, dollars spent)? For categorical targets, the usual test of significance is the chi-squared test (denoted X^2). For numeric targets, the usual test of significance is the t-test.

Figure 3.6: Illustration of a Test of Significance between Means

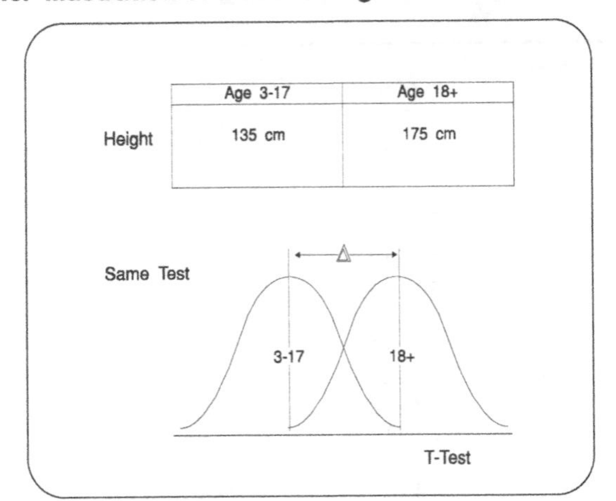

Tuning the Level of Significance

The level of significance affects whether the test succeeds or fails. Different branch targets are produced if different levels of significance are used to perform the test to establish the groupings. Statisticians have evolved rules for selecting the level of significance to use to perform a test. The .05 level of significance is an example. According to this test, the categories that are being compared are collapsed together if they cannot be shown to be significantly related to the value of the target at the 95% level of statistical confidence.[1]

Many other levels of significance can be used—for example, the .01 level (a more conservative test) or the .10 level (a more liberal test). In practice, as more conservative tests are applied to the construction of groups, greater differences between individual levels of encoding need to be observed for these codes to be considered for forming a separate leaf. The net effect is that the selection of the level of significance affects the bushiness of the decision tree and the homogeneity of clusters (in CHAID methods only). At higher levels of statistical significance, it can be hard to reject the null hypothesis of no differences between distributions (i.e., there is not enough separation between their distributions); therefore, many codes are collapsed together. This results in a smaller number of leaves, illustrated in Figure 3.7.

Figure 3.7: Illustration of Branch Granularity as a Function of Cluster Similarity Measures

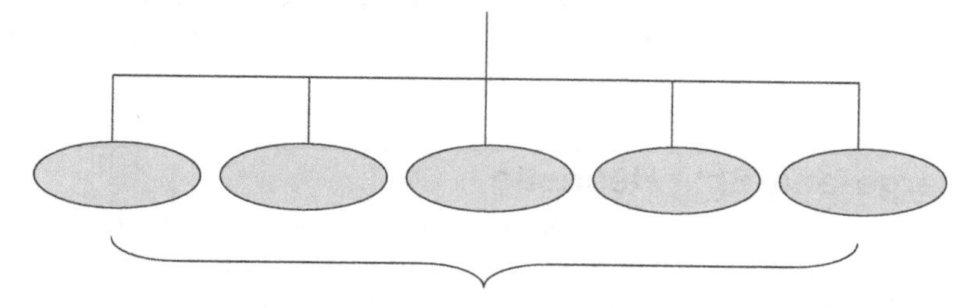

Fine grain (produced with higher levels of significant differences)

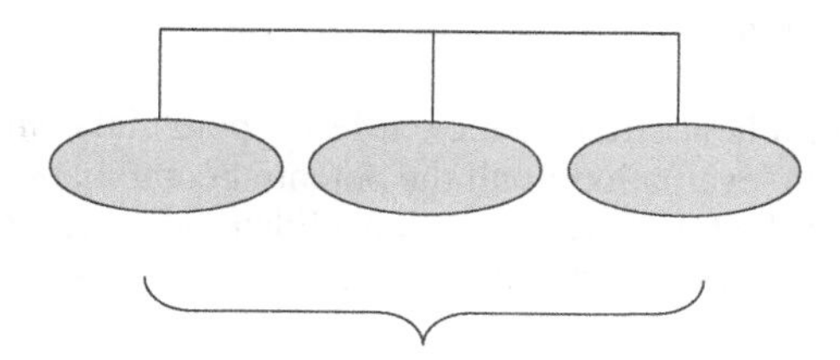

Coarser grain (produced with lower levels of significant differences)

Observation Selection

As the decision tree is grown, successive samples of data set values are used to form the branches of the decision tree. Tree-growing settings determine the number of observations (records) that are taken to form the sample. In the simplest case, all observations are taken to form the sample. However, it is not necessary to take all observations (for example, if the host data set is large). A sample containing less than all observations in the data set can yield identical or equivalent results to the full data set. When samples are taken, it is normal to take independent samples at each node of the decision tree. For nominal targets, it is normal to try to balance the sizes of the outcome categories. For example, suppose a node contains 100 observations of one value of a binary target, and 1,000 observations of another value. If the sample size is set to 200 or more, it makes sense to take all 100 observations of the first target, and to take a random sampling of the other target, until the sample of 200 observations is created. In calculating binary splits, the best binary partition of binary and interval targets is always found.

The creation of multi-branch decision trees is more complicated because of the numerous potential splits (compared to simple, binary splits). You should first consolidate the data before applying the method to evaluate all potential splits or, if a large number of potential splits seem likely, you should use a heuristic search for the best split. A consolidation phase searches for values of the input that would likely be assigned to the same group or cluster in the best split. Simple clustering can be used for consolidation (for example, group all input values that have the same or similar target value). The split-search algorithm treats observations in the same group as if they have the same input value. This results in a faster split search because fewer candidate values need evaluating.

The Kass Merge-and-Split Heuristic

In the development of the CHAID algorithm, Kass specified a merge-and-split heuristic to develop multi-branch trees (Kass, 1980). The merge-and-split heuristic tries to converge on a single, optimal clustering of like codes. This heuristic begins by merging codes within clusters and reassigning consolidated groups of observations to different branches. Then, the merge-and-split heuristic operates as a consolidation algorithm—the consolidated groups are broken up (by splitting out the members with the weakest relationships). These broken-up groups are remerged with consolidated groups that are similar.

The effect of the merge-and-split heuristic is to look at fewer potential combinations of values than would be required by a complete evaluation of all the potential combinations. The process stops when either a binary split is reached or there are no consolidated groups that can be split and merged at the similarity level specified by the algorithm.

SAS Enterprise Miner uses a variation of this heuristic called merge-and-shuffle. The merge-and-shuffle algorithm begins by assigning each consolidated group of observations to a different node. At each merge, the two nodes that degrade the worth of the split the least are merged. After two nodes are merged, the algorithm considers reassigning consolidated groups of observations to different nodes. Each consolidated group is considered and the process stops when no consolidated group can be reassigned.

When using the chi-squared test and F-test criteria, the p-value of the selected split on an input is subjected to more adjustments: if the adjusted p-value is greater than or equal to the worth value, the split is rejected.

Although the merge-and-split heuristic developed by Kass is designed to find a single solution, Biggs et al. (1991) realized that all the intermediate products that are formed in the merge-and-split process can be stored and subsequently evaluated for worth. Then, from all the candidate splits that are stored, the one split with the best worth can be chosen.

Dealing with Missing Data and Missing Inputs in Decision Trees

When forming groups from the values of the inputs, it is common for a data record to contain a missing value. This is almost always true in live data sets, regardless of the amount of data quality and data scrubbing. The net effect is that if the target or input value is missing, it is usually ignored.

In a multivariable technique like decision trees, missing values can lead to a considerable loss of data; once a data record is dropped at any stage of the decision tree growth process, all other data that is available in the data record is lost. For example, a missing value at the top level of the decision tree will cause the data record to be dropped, as well as any other input fields.

Clearly, you want to recover as much of the data as possible in a data record. A number of methods for dealing with missing values in decision trees have been developed.

- treat a missing value as a legitimate value (i.e., explicitly include it in the analysis)
- use surrogates (i.e., another input) to populate descendent nodes where the input value for the preferred input is missing
- estimate the missing value based on non-missing inputs (i.e., treat the missing value category as a target value that can be estimated and, in a two-stage process, include the estimated value in the analysis—the simplest form of this method is to estimate the missing value as the average value for the input)
- distribute the missing value in the input to the descendent node based on a distribution rule (i.e., distribute the missing value to the most common descendent node)
- distribute missing values over all branches in proportion to the missing values by branch

During the search for a split, it is possible to use a missing value as another value when calculating the worth of a split. One advantage of using missing values during the search for splits is that the calculation of the worth of a split is computed with a larger number of observations for each potential split. Another advantage is that, even if there are missing values, this information can increase the predictive accuracy of a split. For splits on a categorical variable, this concept is the same as treating a missing value as a separate category. For continuous (numerical) target splits, this is the same as treating missing values as having the same (unknown) value.

This approach was developed in the original CHAID. A statistical test considers the missing value as another code that is grouped with the class that it most closely resembles (or, the missing values can be grouped into a separate class of their own). There are three variants to accommodate categorical (ordered and unordered) and continuous inputs (ordered, unordered, and floating variations).

An approach pioneered by CRT is to use surrogate splits when there is a missing value for the preferred branch on the decision tree. The input variable to form the split can be missing for an input data record. Other input variables are available (maybe not as strong, but still good) that can be used to determine whether that row of data (observation) goes to the right or the left of the node that is being split. This alternative input variable or surrogate is used to determine where the missing data record is assigned in the descendent node. Both surrogate and competing input variables are alternatives to the input variable that has been selected to form the splits that determine the descendent nodes. Surrogate splits are used only to distribute parent data records to descendent nodes when the selected input variable has missing values that prevent the distribution of records to nodes based on input values. In this case, the surrogate record values determine where to distribute data records in descendent nodes.

Imputation has long been used as a method to handle missing values. The distribution of the valid values for a field in the data set can be defined as a function of a set of inputs in the decision tree or regression form. This means that a predictive equation in the decision tree or regression form is available to produce a score for any target on any record as a function of other fields or inputs in the record. Thus, the values in any field can be considered a function of the values in all the fields in the same data row or record. The predictive equation in the decision tree or regression form can make predictions about unknown situations. A missing value is an unknown situation, so it can be predicted or imputed using this method.

The rules to distribute missing values to descendents in SAS Enterprise Miner are the following:

- distribute missing values across all available branches
- assign missing values to the most correlated branch
- assign missing values to the largest branch

In the distribution approach, data records are distributed to branches in proportion to the size of the branch. Thus, a branch with 50% of the observations, based on valid values, would receive 50% of the data records that contain missing values. (The data records are selected at random.) This method preserves all of the available information and reflects that information in proportion to the size of the branch that it is associated with. This concept is similar to substituting the average value for the decision tree analysis variables whereby the average value is weighted according to the probability of occurrence.

Surrogate Splits

When a split is applied for an observation with a missing value, it is possible to look for surrogate splits on another value before assigning the observation to the branch with the missing values. This surrogate-splitting rule is a backup for the main splitting rule. For example, the main splitting rule uses county. The surrogate-splitting rule uses region. If the county is unknown for a given observation, then region is used in its place.

If several surrogates exist, then when an observation for the main splitting rule is missing, each surrogate is considered in sequence until one can be applied to the observation. If no surrogate can be applied, the main splitting rule assigns the observation to the branch with the missing values.

The surrogates are considered in the order of their agreement with the main splitting rule. The agreement is measured as the proportion of training observations that it and the main splitting rule assign to the same branch. The surrogate rules saved in the decision tree run options to determine the number of surrogates that will be sought. A surrogate is discarded if it has a low agreement with the main splitting rule. A low agreement is less than or equal to 1, divided by the number of branches in the main split. As a result, a node might have fewer surrogates than the number specified in the surrogate rules saved in each node option.

Other characteristics of the calculation of the agreement can be noted. The agreement measure applies only to observations that are valid for the main splitting rule. Of these observations, any instances where the surrogate rule cannot be applied count as observations that do not get assigned to the same branch as the main splitting rule. Thus, an observation with a missing value in the observation used in the surrogate rule but not in the observation used in the main splitting rule counts against the surrogate.

Step 5—Select the Candidate Decision Tree Branches

Once the clustering is complete for all inputs that are being considered as branches at a level of the decision tree, then the inputs can be arranged in a list and ranked according to their predictive or classification power. The measure of power depends on the splitting criterion. In the previous illustration of the calculation of entropy, worth is calculated as the sum of the node computations across a branch. This calculation is used for entropy, Gini, and variance reduction. The test statistic is weighted by the proportion of observations contained in any node of the branch. Prior probabilities can be specified, and if these prior probabilities are incorporated in the split search, then the proportions are modified accordingly.

The chi-squared test and F-test criteria use the worth measure to assess the split. Worth is derived from the traditional p-value that is calculated for these test statistics, and is computed by taking the -log of the p-value. For the test criteria, the best split is the one with the smallest p-value (highest worth). The threshold value (p-value)—used to determine the significance of the test statistic and called the alpha level, which corresponds to the probability of a type I error—is set to .20 by default. The .20 level is liberal by most academic standards (where alpha levels of .05, .01, and even .001 are commonly used), but is considered appropriate in exploratory data mining work. Results need to be confirmed through validation or test trials.

The p-values can be adjusted to account for multiple testing. An approach that follows the original work of Kass is usually used. These adjustments to p-values can be reflected in the display if the adjustment option has been selected. If the Kass adjustment is applied before the split is selected, then the best split is the one with the smallest Kass-adjusted p-value. For nodes with many observations, the algorithm can use a sample for the split search, for computing the worth, and for observing the limit on the minimum size of a branch.

Adjustments can be applied after the split is selected. In this case, the unadjusted worth value is used to select the split. Split worth statistics, shown in the display, can use the Kass-adjusted p-values. Because the Kass adjustments reflect the level of measurement and the type of search, it is believed that the relative worth of inputs is more correctly reflected by adjusted worth measures, and the analyst is less likely to be deceived by the apparent value of a split.

The CHAID Approach

In the CHAID approach, it is common to apply a test of significance to the tables that are formed by each clustered input. The outcome of that test is then used as a measure of the quality of the branch that is formed by that input. The inputs are presented as partition candidates to form the branch of the decision tree at that level of the decision tree growth, in order of statistical confidence based on the test of significance. Statistical adjustments are applied to variables with many categories—ordered and unordered variables are compared on the basis of an adjusted metric so that the metric can be equitably applied to all inputs, regardless of the form of input.

Figure 3.8: Illustration of an F-Test on Multiple Means

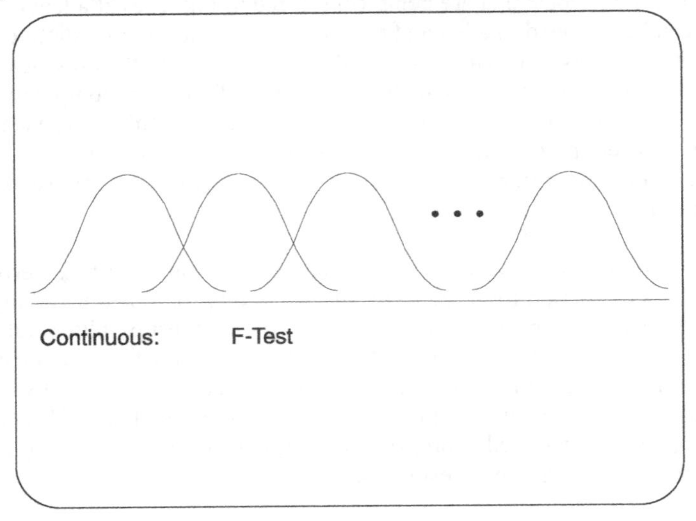

For numeric targets with interval-level measurements, the usual test of significance applied to the candidate branches of the decision tree is the F-test. The F-test provides a measure of between-group similarity versus within-group similarity. Similarity is assessed by measuring the variability of scores around the mean or average values in a descendent node, compared to the variability between the average values across the nodes of a branch.

The F-test calculation is a ratio of the between-node variability (sum of squares) versus the within-node variability. The degrees of freedom are ***n*** (the number of observations) and ***B*** (the number of branches). As shown in Figure 3.9, a ratio is calculated and assessed for significance to determine a p-value or test metric. The significance test uses measures to reflect statistical degrees of freedom, which provides a common way to look at numerical comparisons across different numbers of nodes on a branch and different numbers of observations.

Figure 3.9: Illustration of the Calculation of the F-Test

$$F = \left(\frac{\mathrm{SS}_{\text{between}}}{\mathrm{SS}_{\text{within}}}\right)\left(\frac{n-B}{B-1}\right) \sim F_{B-1,\,n-B}$$

The expression for computing the sum of squares between and the sum of squares within are as follows.

Figure 3.10: Illustration of the Calculation of Sum of Squares Between

$$\mathrm{SS}_{\text{between}} = \sum_{i=1}^{B} n_i \left(\bar{y}_{i\cdot} - \bar{y}_{\cdot\cdot}\right)^2$$

Figure 3.11: Illustration of the Calculation of Sum of Squares Within

$$\mathrm{SS}_{within} = \sum_{i=1}^{B} SS_i = \sum_{i=1}^{B}\sum_{j=1}^{n_i}(y_{ij} - \bar{y}_i)^2$$

In the context of a decision tree, the F-test statistic can be viewed as a measure of deviation of the child leaves of a split compared to the parent, as a function of the pooled variability within the child leaves. The F-test statistic can be used to find branches that have nodes that are distinct from one another, and that have node members that are as homogeneous as possible.

For categorical targets, the usual test of significance applied to the candidate splits is the X^2 (chi-squared test). This test examines the cells of a table, looking for disproportionate numbers of observations in the cells. This happens when greater or fewer observations occur in the cell than would be expected if the observations were distributed randomly. The value of the test statistic increases as more observations collect in one or more cells in disproportionate numbers.[2] In Table 3.1, the relationship between gender and car ownership is shown. Approximately 36% of females (3,606) own a new car, while about 38% of males (1,939) own a new car.

Table 3.1: Relationship between Gender and Car Ownership

Gender and New Car Ownership			
Owns new car	gender		Total
	female	male	
No	6296	3117	9413
(percent)	63.58	61.65	
Yes	3606	1939	5545
(percent)	36.42	38.35	
Total	9902	5056	14958

If there were no relationship between new car ownership and gender, then both males and females would have a 37% rate. This is shown as Observed versus Expected columns for females and males, respectively, in the following table.

```
          Observed                        Expected

          6296        3117                6231        3182
          3606        1939                3671        1874
```

Chi-Squared Test

The chi-squared test is based on calculating the sum of expected, minus the observed frequencies for each cell of the table. These quantities are squared to eliminate negative numbers.

Figure 3.12: Illustration of the Calculation of the Chi-Squared Test

$$\sum \frac{(o-e)^2}{E}$$

The statistic is calculated with respect to the degrees of freedom. The calculation is the number of rows (minus 1), multiplied by the number of columns (minus 1).

Degrees of freedom = (r–1) (c–1); in this case it is 1.

$$X_i^2 = \sum_{i=1,n} \frac{(o-e)^2}{E} = 5.36$$

The table is a two-dimensional table (a crosstabulation table) that shows the distribution of new car ownership within categories of gender. There are a total of 14,958 observations in this data set. A chi-squared value of 5.36 yields a probability of .02 (based on the probability table of chi-squared values). At this point, you might accept the hypothesis that there is no significant difference between males and females if you were using a .01 level of significance. If you were using a .05 level of significance, you would reject the hypothesis. In this case, gender might emerge as a significant input in a decision tree on new car ownership. If you rejected the null hypothesis, then the categories of male and female could not be merged because merging would be treating male and female categories as equivalent.

In a close case such as this, business rules that are derived from business knowledge are often used to determine whether a split is used. Alternatively, a validation sample can be used to determine whether differences between males and females persist, even if the absolute size of the difference is small.

Statistical Adjustments and the Number of Tests

As designed by Kass, the CHAID algorithm provided a method to apply statistics to assess the quality of the branches that were selected for presentation on the decision tree. Part of the statistical-testing framework included adjusting the level of significance to accommodate the number of tests of significance that were applied to determine the characteristics of a branch (statistical tests are used in forming the clusters). It is common in statistical hypothesis testing to adjust the level of significance according to the number of tests applied to a sample. This is because tables of statistical significance have been prepared assuming there is one test on one sample of data. The adjustments help prevent overfitting based on a calculation of a test statistic that is overstated.

These statistical adjustments—called Bonferroni adjustments—are designed to return a true probability level for statistical confidence that is independent of the number of statistical tests that formed the branches of the decision tree. In Figure 3.13, more statistical tests are performed with unordered categories in branch clustering than with ordered categories. Adjustments that consider the number of tests that form the branches regard either method on the same basis in terms of the values of the computed probability levels.

Figure 3.13: Illustration of Computing the Number of Statistical Tests

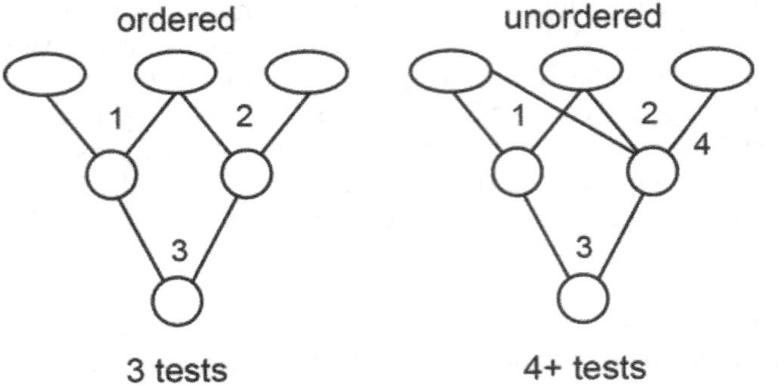

Branches are reported with a true level of significance or a worth metric; furthermore, all branches are reported on the same basis. This means that branches that consist of many values or branches that are unordered (and therefore can combine in many ways) are appropriately adjusted so that they can be evaluated on the same basis as a branch with fewer categories. The net result is that alternative splits are presented in a valid order in terms of their probability levels.

Research by Biggs et al. (1991) led to the development of the exhaustive method of identifying branch splits. This method incorporates Bonferroni adjustments that should be applied for the various types of inputs and number of categories that are included in the analysis. (Original work by Kass suggested Bonferroni adjustments that were more conservative than necessary.)

Other adjustments have been incorporated into the SAS Enterprise Miner decision tree. The Kass adjustment (1980) can cause the p-value to become unnecessarily more conservative than an alternative method, called Gabriel's adjustment, does. In this case, Gabriel's p-value is used.

A depth adjustment can adjust the final p-value for a partition to simultaneously accept all previous partitions used to create the current subset being partitioned. The CHAID algorithm has a Bonferroni adjustment within each node, but it does not provide a multiplicity adjustment for the number of leaves. For example, imagine an extreme case where a decision tree has grown to a thousand leaves. If a significance test were conducted in each leaf at an alpha level of 0.05, a CHAID algorithm would obtain about 50 false test-of-significance outcomes (reject the null hypothesis of no differences between two leaves in a decision tree). Hence, the decision tree is likely to grow too big. The depth receives a Bonferroni adjustment for the number of leaves to correct the excessive number of rejections.

In addition, there is a method to adjust the p-value for the effective number of inputs. The more inputs, the more likely an input will accidentally win over the truly predictive inputs. The more correlated inputs, the more likely the risk. The input adjustment multiplies the p-value by the number that is declared for the effective number of inputs. The default effective number of inputs equals the number of inputs that are declared live in the analysis.

An Example

Data is in the form of Amount Purchased, including Time of Purchase, Quantity Purchased, Age of Customer, and Distance Traveled. Quantity Purchased is the target field. The task is to describe the target field—**Quantity**—in terms of the other input fields.

Age	Date	Hour	Distance	Quantity	Amount	Category
35	3/21/2003	6	2	2	14.95	Shelving
29	3/21/2003	6	5	2	29.9	Shelving
40	3/21/2003	7	9	5	39.8	KitchenWare
33	3/21/2003	7	44	5	12.71	KitchenWare
50	3/21/2003	8	33	5	37.35	Shelving
27	3/21/2003	9	8	5	20	Shelving
34	11/11/2003	9	10	1	78.6	Bathrooms
58	5/17/2002	1	37	9	78.37	Bathrooms
37	5/17/2002	2	22	9	39.95	Electrical
39	5/17/2002	2	12	9	34.9	Books
24	5/17/2002	3	7	9	73	Bathrooms
44	5/17/2002	3	51	9	14.95	Bathrooms
41	5/17/2002	4	6	9	78.6	Music
30	5/17/2002	6	1	9	20	Bathrooms

With **Quantity** as a target, potential inputs include **Age**, **Date, Hour**, **Distance**, and **Category**. Total Amount could be used as an input, but would usually not be because the two measures are tightly related. In many software products, the user does not usually control the search order of inputs. In this example, the software begins by looking at the association between the target **Quantity** and the time input of **Hour**.

Some preprocessing is required before the decision tree is grown.

1. Continuous inputs need to be converted to categories.
2. The search order of inputs needs to be determined. Is the input categorical or continuous?
3. The number of allowable branches needs to be determined.
4. The similarity measure, used to combine similar categories, needs to be determined.

After this preprocessing, the target **Quantity** is set to be modeled as determined by the inputs **Age**, **Date**, **Hour**, **Distance**, and **Category**. **Hour** and **Distance** have been calculated so that the categories are meaningful. The decision tree algorithm begins by examining time, looking through such combinations as:

Hour combinations

9-10
9-10-11
9-10-11-12
9-10 vs. 11-12

and so on.

The goal of this step is to find a meaningful combination of input values that can usefully describe variations in the Quantity Purchased. This could produce a decision tree like Figure 3.14.

Figure 3.14: Illustration of a Decision Tree of Quantity Purchased Grouped by Hour

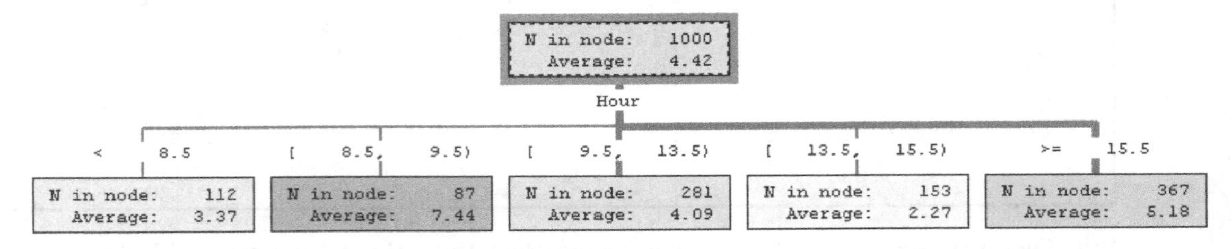

It could also be a cube-like dimensional representation like Figure 3.15.

Figure 3.15: Illustration of a Translation of Decision Tree Results to a Cube Display

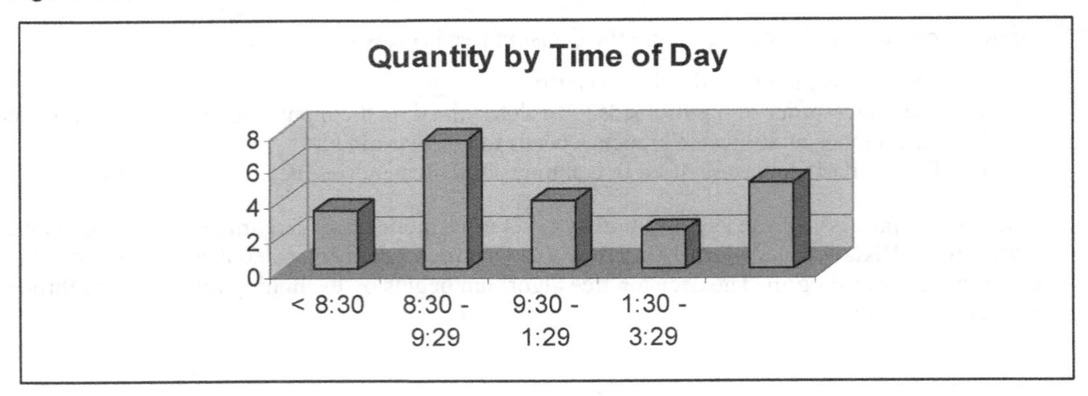

As you can see, there is a direct relationship between the branches of a decision tree and the cells of a table. The notations that describe the branches of the decision tree indicate the range of acceptable values; for example, (9.5–13.5) indicates a range of values from 9.5 (including .5) to 13.5 (but not including .5), the last value before 13.5 (this 24-hour clock time is shown as 1:29 in the 12 hour clock display shown in Figure 3.15).

The decision of whether to combine values is made by a numerical or statistical test—essentially, these tests combine codes that are alike (with respect to the target), while distinguishing them from other codes.

Using the previous example, assume the next input to be evaluated is **Age**. The decision tree algorithm tries to find the best way of characterizing **Quantity** as a function of **Age**. Assume that the first age category was 14, and that the combined value for **Quantity** in the 14 age category was 5. Assuming that the next age category is 15, the decision tree algorithm essentially sets up a test of similarity between the quantities in the age category 14 compared to the quantities in the age category of 15.

The decision tree algorithm examines 14 and 15. If the two categories are similar, then the categories are combined. If age categories 14 and 15 are combined, then this combined category is compared to the distribution of target values in the age category of 16.

14+15+16 same?

If yes, combine, and so on.

14+15+16 vs. 17+18+19

Maybe the best age profile is what is shown in Figure 3.16.

Figure 3.16: Illustration of Branch Partitions Applied to a Dimensional Display

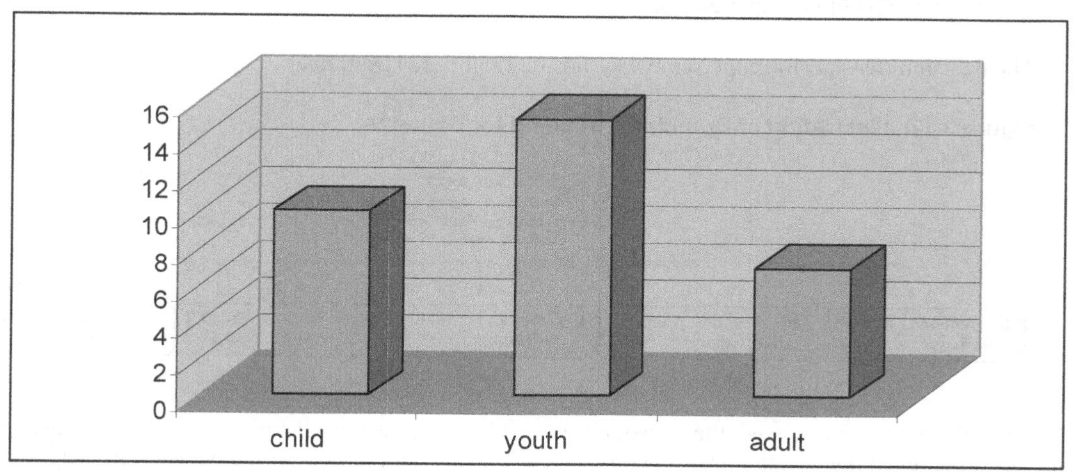

Several things are going on here:

1. This is a continuous target. If the target is categorical, the process is similar. However, the test of similarity is different. With continuous targets, the test of similarity compares variance around the average in each of the groups that are formed by the input categories. With categorical targets, a test of significance (typically, a chi-squared test) or a Gini or entropy test can be applied.
2. Multi-branch trees (i.e., more than 2 leaves) are allowed.
3. Only monotonic combinations are being looked at (i.e., combinations of a lower-valued quantity, such as 3, with a higher-valued quantity, such as 4).
4. The specifics of the test determining whether two categories are the same are not being discussed. Typically, a test of significance is used. It is possible to indicate ahead of time that you want 3-way branches or 5-way branches. In this example, the decision tree algorithm tries to split the categories into the specified number of branches to maximize the inter-branch category values and to force the greatest amount of intra-branch differences.

There are several splitting criteria. For interval targets, there are the following:

- variance reduction
- F-test

For nominal (categorical) targets, there are the following:

- Gini or entropy reduction (information gain)
- CHAID or chi-squared test

The decision trees perform the same way for the final field (**Distance**).

Figure 3.17: Illustration of Branch Partitioning for Distance

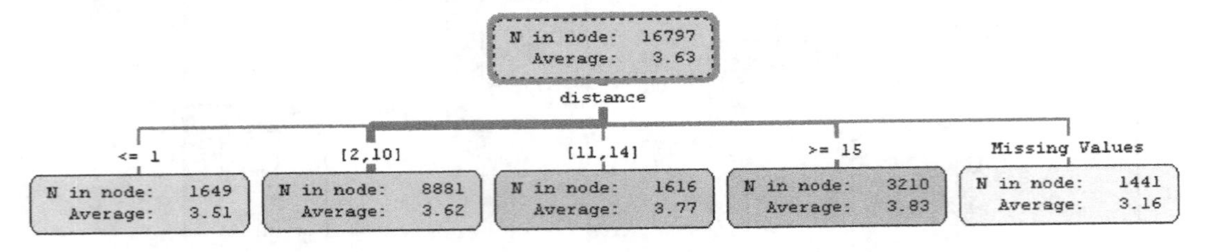

After this step is completed, the decision tree will have a candidate list of branches that could form the branches of the decision tree. If you could see inside the memory of the decision tree algorithm, you might see a table that looks like the following:

Figure 3.18: Illustration of the Candidate List for Node Partitioning

Once the values are combined, alternative branches can be compared to determine how strongly they relate to the target. This information is often used to select the appropriate branch to form the first level of the decision tree.

In the interactive mode of operation in the SAS Enterprise Miner decision tree, this candidate list of branches is displayed for selection. If you were to browse the list in interactive mode, a set of displays as shown in Figure 3.19 might be produced.

Figure 3.19: Illustration of Potential Branches in Interactive Mode

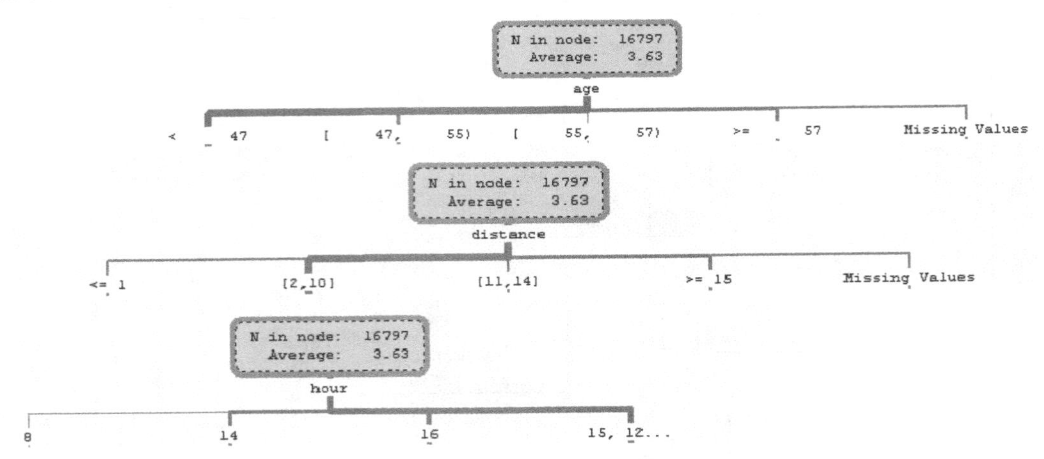

The decision tree algorithm inspects each one of these views to identify which view to choose as the splitting criterion to form the decision tree. Assume that the algorithm selected **Age** as the splitting criterion. This would produce a decision tree such as the following:

Figure 3.20: Final Branch Partition Selected at This Level of the Decision Tree

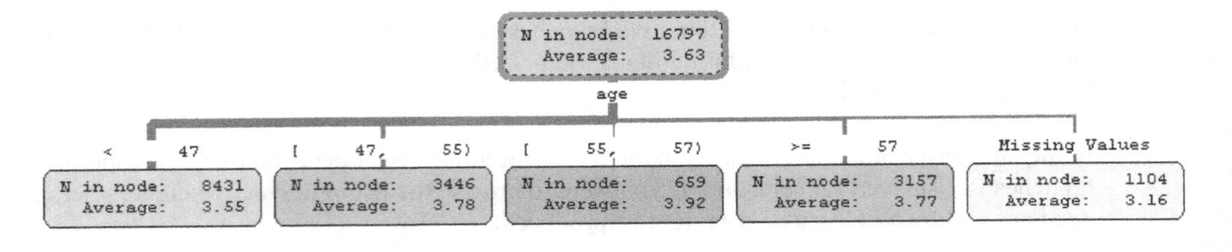

In the original CHAID algorithm, the inputs are either nominal or ordinal. Most software applications accept interval inputs and automatically group the input values into categorical ranges of discrete nominal or ordinal values before growing the decision tree. The splitting criteria are based on p-values from the F-test distribution (interval targets) or chi-squared test distribution (nominal targets). The p-values are adjusted to accommodate multiple testing. A missing value can be treated as a separate value. For nominal inputs, a missing value constitutes a new category. For ordinal inputs, a missing value can group with the code grouping it most closely resembles (as determined by a test of significance).

The search for a split on an input proceeds step by step. Initially, a node is allocated for each value of the input. Nodes are alternately merged and re-split, according to the intra-branch p-values. The original CHAID algorithm stops when no merge or re-splitting operation creates an adequate p-value. Then, the final split is adopted. A common alternative to this split method, called the exhaustive method, continues merging to a binary split, and then adopts the split with the most favorable p-value among all of the splits that the algorithm considered. After a split is adopted, its p-value is adjusted, and the input with the smallest adjusted p-value is selected as the splitting variable. If the adjusted p-value is under the threshold you specified, then the node is split. Decision tree construction ends when all of the adjusted p-values of the splitting variables in the unsplit nodes are above the user-specified threshold (because they do not meet the test of significance).

The SAS Enterprise Miner decision tree implementation is different from the original CHAID algorithm.

- CHAID transforms interval inputs into discrete categories (bins or deciles); the SAS Enterprise Miner decision tree consolidates observations into groups. In the original CHAID algorithm, these transformed interval inputs were maintained throughout the analysis. The SAS Enterprise Miner decision tree can group and regroup interval and ordinal values dynamically as the decision tree partitions data throughout the analysis.
- The decision tree node searches on a within-node sample, unlike CHAID.

The CRT Approach

In the CRT approach, it is common to grow a decision tree with more branches and sub-branches than the CHAID approach grows. The CRT approach relies on pruning to cut the branches that do not perform well. Whereas performance in CHAID is determined by a test of significance, in CRT, performance is determined by a validation approach. The CHAID approach grows decision trees with more than 2 nodes in the branches; the CRT approach grows only 2-way (binary) branches.

In CRT, branches can be selected by the following:

- Number of Leaves. When this selection method is used, the branch with *n* leaves (where *n* is the number of leaves selected in the user interface) is selected. Leaves refer to the number of terminal nodes on a decision tree.
- Best Assessment Value. This method chooses the smallest branch with the best assessment value. The assessment is based on validation data, when available.
- The Most Leaves. This method chooses the largest branch after pruning nodes that do not increase the assessment (based on training data).
- Gini (CRT). Gini is used in the CRT method. CRT uses probabilities to compute the impurity of the nodes. The formula for a node ***t*** is computed as:

  $$i(t) = 1 - S$$

  S is the sum of the squared probabilities of the components of the node. Impurity is a measure of homogeneity in the node membership for classification decision trees.

- Variance Reduction. This method is used to compute the best assessment value of a branch when interval data is used for the target.

For CRT, the inputs are either nominal or interval. Ordinal inputs are treated as interval inputs. The traditional splitting criteria are the following:

- for interval targets, variance reduction and least-absolute-deviation reduction.
- for nominal targets, Gini and impurity.
- for binary targets, Gini, Twoing, and ordered Twoing create the same splits. Twoing and ordered Twoing are used infrequently with binary targets. These criteria are considered superior to entropy or Gini criteria with multi-valued discrete targets.

The CRT method does an exhaustive search for the best binary split. Linear combination splits are also available. Using a linear combination split, an observation is assigned to the left branch when a linear combination of interval inputs is less than a specified constant. The coefficients and the constant define the split. The CRT method for searching for linear combination splits is heuristic, and might not find the best linear combination.

When creating a split, observations with a missing value in the splitting variable (or variables, in the case of linear combination) are omitted. Surrogate splits are created and used to assign observations to branches when the main splitting variable is missing. If missing values prevent the use of the main and surrogate splitting variables, then the observation is assigned to the largest branch (based on the within-node training sample).

When a node contains many training observations, a sample is used for the split search. The samples in different nodes are independent. For nominal targets, prior probabilities and misclassification costs are recognized.

The decision tree is purposefully grown to contain branches and subtrees that are not stable from the point of view of reproducibility in a new data set (or by reference to a validation data set). This intention is called overfitting. A sequence of subtrees is formed at each split. The splitting criteria are based on a measure that includes maximum-divided-by-minimum node size and the depth of the decision tree. These three measures—maximum, minimum, and depth—are used as measures of complexity. The assessment measure is calculated and used to construct each subtree. Accuracy is used as the assessment measure. If a profit matrix is available, then profitability can be used as an assessment measure. Accuracy can be computed based on a training sample, a validation data set, and a cross-validation approach.

For nominal targets, class probability decision trees are sometimes used as an alternative to classification trees. Decision trees are grown to produce discriminations between the distributions of class probabilities in the leaves. Decision trees are evaluated by the overall Gini index.

Retrospective Pruning, Cost-Complexity Pruning, and Reduced-Error Pruning

Retrospective pruning originated with cost-complexity pruning and is described in the development of the CRT algorithm by Breiman et al. (1984). This pruning method attempts to identify the best subtree. The "best" is determined by predictive accuracy, weighted by the number of leaves in the subtree. This method is a kind of Occam's razor, meaning that the subtree with the highest accuracy and fewest leaves is chosen over any other subtree that has a similar predictive accuracy.

The decision tree in SAS Enterprise Miner provides the ability to create subtree sequences, using either the training data or validation data to compute the assessment values for choosing subtrees in the sequence. Using the training data produces a sequence that would result from using cost-complexity pruning, which was developed by Breiman et al. Using the validation data produces a sequence that would result from using reduced-error pruning, as described by Quinlan (1987). Reduced-error pruning relies exclusively on validation data; it finds the subtree that is best for a validation data set and does not rely on the creation of sequences of subtrees.

Selecting the Final Branch

The decision of which branch to select to form the split is an important one because the form of the subsequent decision tree depends entirely on which branch, with which number of nodes or leaves, is selected. Following are some of the main considerations:

- Select the branch that will develop the best descriptive model for the analysis.
- Select the branch that will develop the best predictive model for the analysis.
- Select the branch that will develop the best explanatory model for the analysis.

If the goal is to construct a descriptive model, then it is best to create splits in branches that reflect the business user's conceptual approach to the subject area. In this case, the strength of a split on a decision tree is less important than the form of a branch. For example, if states are split into regional groups that reflect the business structure of the enterprise, the description is enhanced. This could be a preferable split over one that yields a better predictive result, but is less intuitive.

When prediction is the goal, the form and shape of the decision tree might never be examined at all. In this case, you want good predictive results. This is measured by the validated prediction or classification rate of the decision tree. The result can be visually inspected using indicators such as a lift chart.

When the goal is explanation, it is useful to grow the branches of the decision tree in a particular sequence. This enables you to construct a decision tree where earlier effects are introduced higher in the decision tree. Therefore, their effects on lower or later effects can be gauged. Sequencing can be used to suggest the form of the interactions among inputs with respect to the target. This is a way of using the decision tree to support explanations that relate to presumed sequences of events and interrelationships. This treatment is not usually possible if the decision tree is grown for simply descriptive purposes or for maximizing prediction.

Step 6—Complete the Form and Content of the Final Decision Tree

After the first level of the decision tree is formed, the decision tree algorithm can be applied recursively to the nodes (or, at this point, leaves) of the first branch. These nodes become candidates for splitting into branches, in the same way as the original root node was examined. This process continues recursively until a full decision tree is grown. The process can be stopped in a number of ways, as discussed below.

It is worthwhile to stop at a good point to avoid overfitting the decision tree. Detecting overfitting is important for the following reasons:

- If the decision tree is overfitting the data, then the relationships that are displayed in the decision tree are not stable and could be a source of misunderstanding about the relationships in the data.
- If the decision tree is overfitting the data, then predictions, which are based on the structure of relationships as identified in the decision tree, will not be good. This means that the predictive power and reproducibility of the decision tree will be weak.

Stop, Grow, Prune, or Iterate

Statistical measures and validation methods can be used to decide how large to grow a decision tree and to evaluate the quality of the decision tree.

After the branch has been selected to form the first level of the decision tree (which is below the root node), the process of splitting is repeated for each of the leaf nodes in the new decision tree to fill out the decision tree to its final form.

The process of forming the CHAID decision tree continues until a node is selected that cannot produce any significant splits below it. Or, the process continues until a stopping rule is encountered. A typical stopping rule might be: "Do not split any node with less than 10 records in it" or "Do not create any node with less than 10 records in it".

The process of forming the CRT decision tree begins much like the CHAID process. CRT forms binary decision trees, rather than multi-way decision trees. While CHAID uses adjusted tests of significance to stop tree growth, CRT relies on validation tests to prune branches, to stop tree growth, and to form an optimal decision tree.

In CHAID, after a node is split, the newly created nodes are considered for splitting. This recursive process ends when no further node can be split. The reasons a node can no longer split are the following:

- The user has decided to stop.
- The node contains too few observations to split in a meaningful way.

- The maximum depth of the decision tree (i.e., the number of nodes in the path between the root node and the given node) exceeds a specific level (typically set by the user).
- No split exceeds the threshold worth requirement specified in the F-test or chi-squared significance level value, or in the variance reduction setting.

The last reason is the most informative. Typically, in this situation all the observations in the node contain almost the same target value, or no input in the node is sufficiently predictive. The decision tree approach is very effective at developing a strong fit between the branches of a decision tree and the data that is used to discover the particular form of the decision tree. However, this type of use comes at a price. The specific form of the decision tree, particularly at lower levels, cannot be exactly reproduced when applied to new data. Decision trees that fit the training data at deeper levels often predict too poorly to apply to new data. While the general form of the higher-level branches might track new data well, lower-level branches are more idiosyncratic and cannot usually reproduce in new data.

When the basic defaults for growing the decision tree are set to extreme values, the decision tree is likely to grow until all observations in a leaf contain the same target value. Such decision trees overfit the training data and will poorly predict new data.

A primary consideration when developing a decision tree for prediction is deciding how large to grow the decision tree or what nodes to prune. The CHAID method specifies a significance level of a chi-squared test to stop tree growth. The originators of the C5 and CRT methods argue that the right thresholds for stopping tree growth are not knowable in advance. They recommend growing a decision tree too large and then pruning nodes.

The SAS Enterprise Miner decision tree node provides both the CHAID approach and the grow-and-prune approach. A sequence of subtrees of the original decision tree is always grown—one subtree for each possible number of leaves. After the sequence of subtrees is established, the decision tree node uses one of four methods to select which subtree to use for prediction:

1. most leaves
2. at most indicated number of leaves
3. best assessment value
4. average profit

The user typically determines the desired subtree method. Options available include:

- if at most indicated number of leaves = n subtree (n is the number of leaves in the subtree)
- best assessment value

If the first approach is selected, then the decision engine uses the largest subtree with at most n leaves. If the second approach is selected, then the decision engine uses the smallest subtree with the best assessment value.

The decision tree stops growing at a certain point, depending on the outcome of this assessment. The assessment is based on the validation data when available. If the subtree method is set to **most leaves**, then the node uses the largest subtree after pruning nodes that do not increase the assessment. For nominal targets, the largest subtree in the sequence might be much smaller than the original unpruned tree because a splitting rule can have a good branch assessment value (split worth) without increasing the number of observations correctly classified.

Assessment Measures

The most common assessment measure is proportion correctly classified if the target is qualitative or categorical, and the sum of squared errors if the target is quantitative. For continuous targets, average square error is used.

Other assessment measures include proportion of event in the top 50% on target 1. This uses the half of the observations that are predicted most likely to equal 1, and uses the training data to compute the proportion in which the target equals 1. This measure can be extended to include proportion of event in the top *x*% on target value *y*. This uses a user-defined threshold as an alternative to 50% to observe the successful classification rates at an arbitrary percent level (for example, 33 to compute the percentage of success in the top third).

In summary, the CHAID method does the following:

- performs subprocesses 2–4 in the selected node, for all nodes
- stops when no more branches are significant
- stops when cell sizes fall below a certain threshold (or when nodes of a certain size cannot be produced)

The CRT method picks the best subtree for each of the extremities of the decision tree through pruning. The final decision tree is the tree that is left after the subtrees have been pruned according to the tree growth selections.

Key Differences between CHAID and CRT

The main difference between a CHAID approach to growing a tree and a CRT approach lies in whether a test of significance or a train-and-test measurement comparison is used.

In the classical CHAID approach, a test of significance forms the groups of codes that form the branch. In turn, this branch is evaluated with a test of significance to determine whether it is used in the decision tree. In the CRT approach, a number of methods can be used to form the branches (although, classically, a variance reduction approach is used to form binary branches). The resulting branches are tested against a validation sample to determine whether the branch accuracy is sufficiently high enough to be used in the decision tree.

Accuracy can be computed many ways. In the simplest way, a decision tree is grown and the predicted classification or prediction is tested against the data set used to train the form of the decision tree. This is called a resubstitution test. The predicted score is substituted for the original score in the training data set, and the overall accuracy rate is computed by comparing the substituted score with the original score.

The resubstituted accuracy rate appears higher than the true accuracy rate because the same data that is used to train the form of the decision tree is used to test the efficacy of the form that was trained. Training the form of the decision tree in this way might pick up idiosyncrasies in the training data that are specific only to the training data. This means that the data is not reflective of the data universe that the training data is designed to reflect. So, the trained decision tree contains these idiosyncrasies. The accuracy rate is computed on training data that includes the idiosyncrasies because this is the same data that was used to train the decision tree.

If a new data set from the data universe was used to test the accuracy of the decision tree, it is unlikely that the new data would include the idiosyncrasies that the trained data included. The accuracy rate that would be computed with this new data is less than the accuracy rate that would be computed using resubstitution.

One-pass resubstitution always overstates accuracy rate. Multi-pass methods such as cross-validation or boosting through re-sampling produce better results. A better, more accurate rate would come from the use of a new, independent data set. This is the preferred method of computing the accuracy rate. This means that a separate data set is made from the original data and is used for testing purposes only. Because it is made before the testing begins, this data is not likely to have idiosyncrasies that will be trained in the decision tree.

Guiding Tree Growth with Costs and Benefits in the Target

When there is a gain and loss associated with correct and incorrect predictive decisions, it is important to incorporate cost and benefit into the selection of decision tree targets. A cost or profit can be assigned to an outcome. The implied profitability of a recommended outcome (the prediction) is used to determine the final form of the decision tree. Psychologists have shown that implied costs and benefits lie behind a wide range of human decision-making. This theory is based on the theory of signal detection (Green and Swets 1996). For example, the decision of whether you have enough gas to get to the next gas station when driving on a road carries a different weight than the decision of whether you have enough gas to get to the next airport when flying a jetliner. On the road, the implied saving of time might easily outweigh the potential cost of running out of gas before arrival; this is not so on the jetliner. Timeliness is important, but easily offset by the potential cost associated with a catastrophic loss of life. Moreover, different decision-makers (as well as their decision-making tools) make different decisions based on their decision-making style (for example, whether they tend to be conservative or more liberal).

Guidance in decision-making tasks is provided by referring to a confusion matrix, as shown in Figure 3.21. The event of interest—**X**—can either occur or not occur. So, it is either **X** or **Not X**. The confusion matrix compares the actual distribution of **X** and **Not X** to the observed (or predicted) distribution. Hits happen when the predicted event—**X**—actually occurs. Misses happen when, for example, it is predicted that there is an **X** event, but, in actuality, it is **Not X**.

Figure 3.21: Confusion Matrix

		Guess/Prediction	
		X	Not X
Actual	Not X	Miss	Accurate Rejection
	X	Hit	False Alarm

Signal detection assesses the effectiveness of the decision boundary that is used to determine whether an event is **X** or **Not X**. The idea of signal detection is shown in Figure 3.22. There are two distributions—A and B. Distribution A represents the distribution of an outcome; for example, the probability that a customer who buys a pair of jeans will also buy an accompanying sweatshirt. Distribution B represents the same distribution with noise or uncertainty added. This noise comes from a variety of sources, such as a conservative versus liberal decision-making style. The difference between the two distributions is illustrated by the line showing the distance between the peaks of the two distributions.

Figure 3.22: Illustration of Signal Versus Noise Distributions in Signal Detection

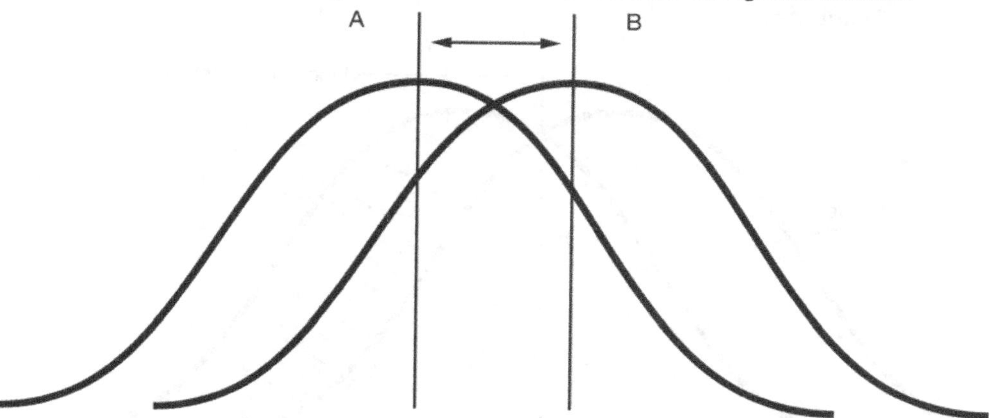

The signal detection theory says that, given the uncertainty represented by the difference between the two distributions, decisions can result in hits, misses, false alarms, and correct rejections. If you guess that the customer will buy and the actual result is a purchase, then you have a hit. If you guess that the customer will buy, but the customer does not buy, then you have a false alarm. If the customer buys and you guess that the customer will not buy, then you have a miss. One goal of signal detection is to determine the ideal circumstances that maximize correct decisions (hits and correct rejections), while minimizing incorrect decisions (misses or false alarms).

This can be illustrated by referencing the area under the distributions in Figure 3.23. You see the effect of a default 50% decision threshold that represents a halfway point between conservative and liberal. The area to the right of the threshold represents hits, and the area to the left represents misses.

Figure 3.23: Illustration of the Effect of Decision Threshold on the Signal Versus Noise Distributions

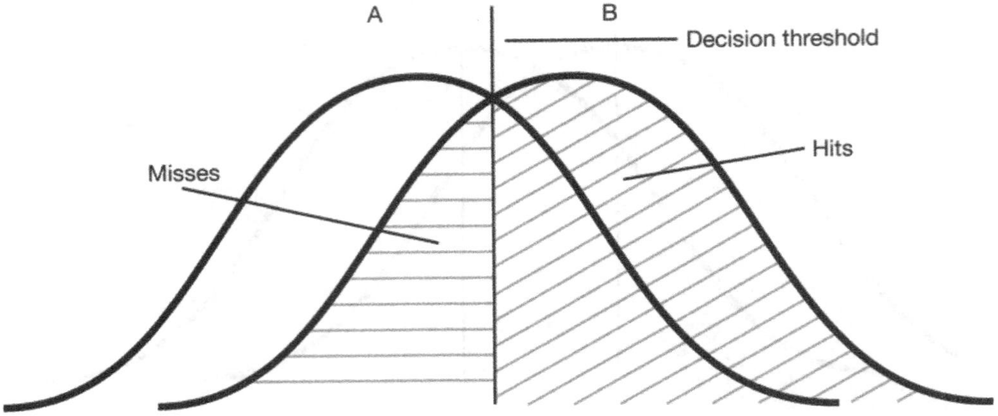

Costs and benefits can be used to construct predictive decision trees that are accurate (regardless of the decision method) and produce the most profitable result. This is possible because decision thresholds are rarely clear-cut and can be changed to reflect costs and benefits.

Figure 3.24: Illustration of Hit Rate Given the Decision Threshold on the Signal Versus Noise Distributions

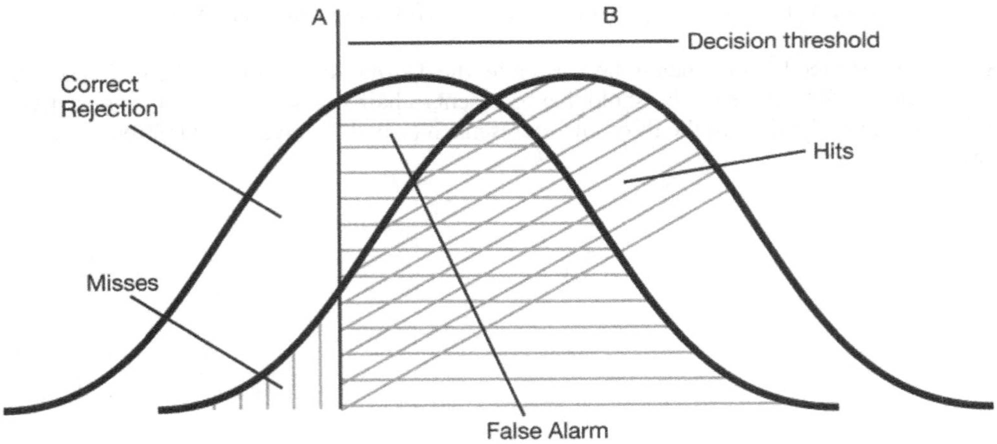

Other Software Features

As shown in Figure 3.25, a change in the decision threshold can change the proportions of hits, misses, and false alarms. When comparing Figure 3.23 to Figure 3.24, you see that while hits increase and misses decrease, there is an accompanying increase in false alarms.

This situation is shown in Figure 3.25. A grid is set up to compare the distribution of values that are predicted by the decision engine at a given threshold. If you predict the presence of an event—designated with **X**—then you have a hit. If there are events in the data, but you do not predict them, then you have a miss. Similarly, when you predict an event, but the event is not there, you have a false alarm.

Figure 3.25: Illustration of Hits and Misses at a 50% Decision Threshold

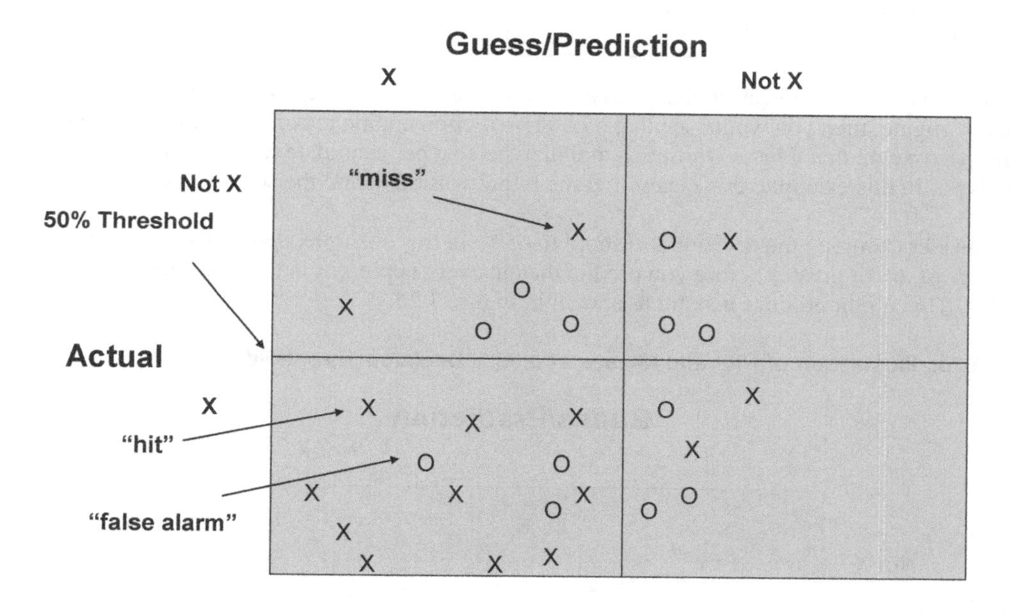

Figure 3.26 shows the distribution of a target value in a record (illustrated as **X**) and whether the value is detected at a baseline threshold. For illustration, assume that the decision threshold is 50%—a common setting for most decision-making tasks. This threshold reflects the notion that if the probability of an event, as estimated by the prediction engine, is .5 or greater, then you will set the decision to the event (in this case, **X**). If the probability is less than .5, then you will set the decision to **Not X**. A results table, called a misclassification table, is shown in the following table. There are 15 **X** events in the prediction space and the prediction engine has correctly identified 10 of them. This provides a sensitivity of about 67%. This sensitivity was gained from a decision threshold of 50%, which returned 13 predicted **X** events overall. The success rate of this prediction engine is 10 out of 13—77%. This is called the specificity.

Table 3.2: Misclassification Table

	Actual		
Predicted	*x*	*not-x*	*sum*
not-x	5	9	14
x	10	3	13
sum	15	12	27

If you assigned a benefit or return metric to a hit, and assigned a penalty or cost to run the prediction engine, then you would see that a good prediction engine maximizes sensitivity and specificity. Assume that a hit is worth $20 and that the cost per candidate to run the prediction engine is $5. In this example, this means that the return was $200 and the search cost was $65.

Now, consider changing the decision threshold to 65%. In this example, demand a predicted probability of .65 or greater before you predict that an event is present in the data records being classified. This might produce a result as shown in Figure 3.26.

Figure 3.26: Illustration of Hits and Misses at a 65% Decision Threshold

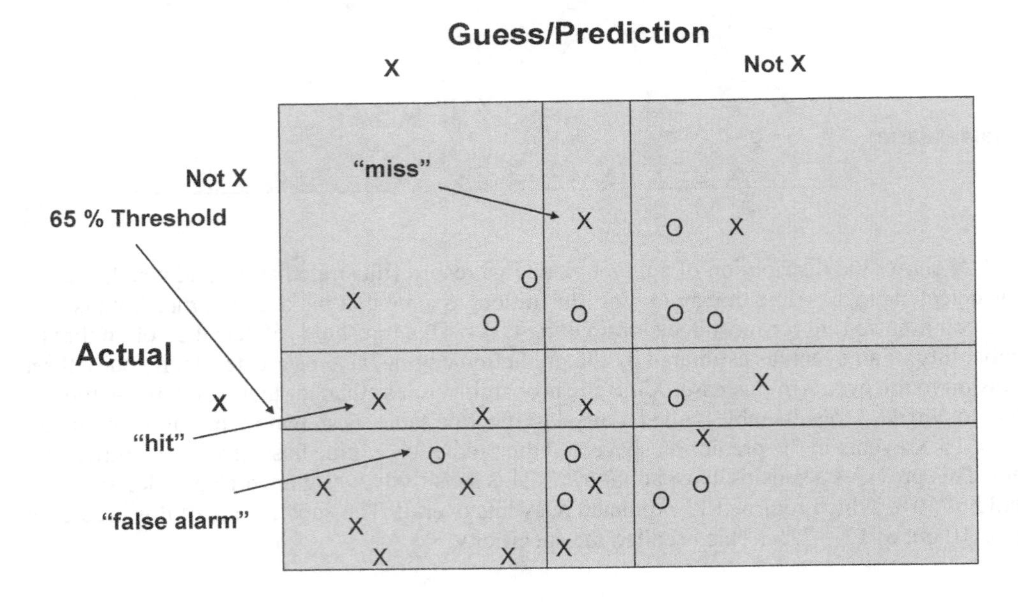

If you look at the return on investment of this decision approach, you see that there were 12 hits (for a return of $240) and that the search cost was $115.

	Actual		
Predicted	*x*	*not-x*	*sum*
not-x	3	1	4
x	12	11	23
sum	15	12	27

Now, look at the two approaches, as shown in the following table:

	50% Threshold	*65% Threshold*
Sensitivity	0.67	0.80
Specificity	0.77	0.52
Return on Investment	200	125

If you use the typical 50% cutoff in estimating the probability of an event (for example, an additional purchase), then you will get *x* hits. You also get *y* misses. Now, if you shift the decision threshold to a 65% cutoff, then you increase the number of hits to *x*+1. The misses decrease, but, as expected, the number of false alarms increases.

Prior Probabilities

Prior probabilities for the target classes can determine whether the counts and proportions in the formula for the assessment measure are adjusted by prior probabilities. For example, suppose 60% of the observations have a target value of 0, and the remaining 40% have a target value of 1. Assume that the decision tree predicts all observations to be 0, and the prior probability of 0 is 10%. Because the decision tree predicts all zeros, then it appears that the misclassification rate is 40% (because the apparent distribution shows 40%). If prior probabilities are incorporated in the assessment measure, then the proportion misclassified would be 90%. Otherwise, it would be 40%.

Figure 3.27: Illustration of the Effect of Prior Probabilities on Apparent and Real Distributions

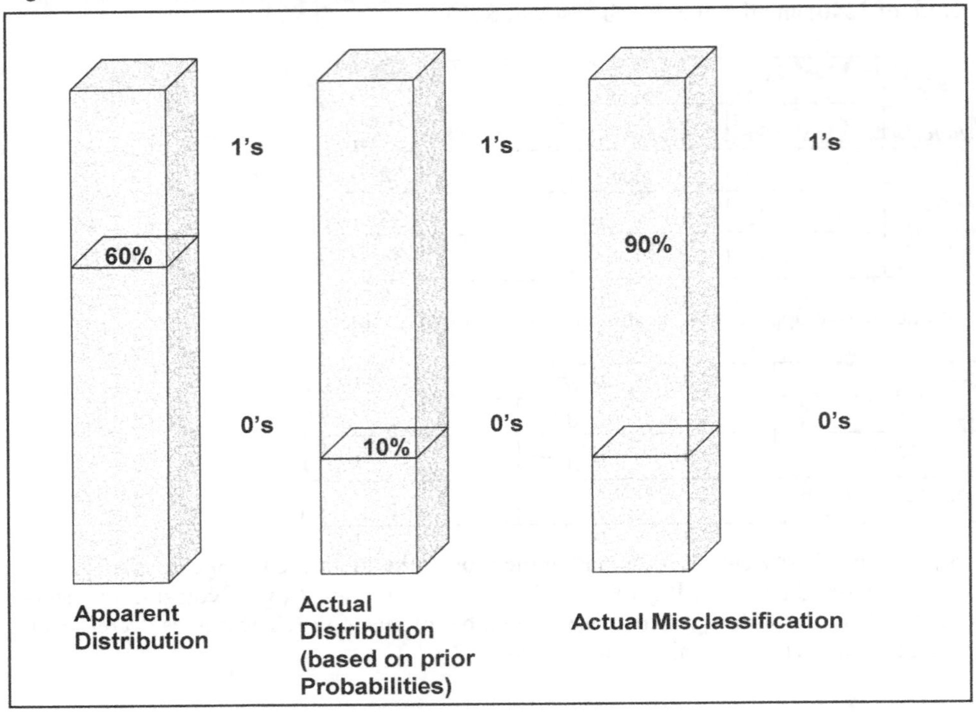

Prior probabilities do not change the shape of the decision tree. The decision tree makes the same prediction, regardless of whether the assessment measure incorporates prior probabilities or not. Only the assessment results change with the prior probability specification.

The decision tree always uses prior probabilities when predicting a target value. If the training and validation data sets are obtained by oversampling observations that have a rare target value, then incorporating prior probabilities in the misclassification rate could offset the goal of oversampling, which would artificially boost the apparent incidence of a rare code in the training data. On the other hand, when the assessment measure is the proportion of the event in the data, then the incorporation of prior probabilities would give a better idea of how the decision tree will perform when it is deployed in a live environment with new data.

Switching Targets

One of the more recent innovations that have been introduced in decision tree modeling is the ability to define multiple targets and switch between targets as the tree growth progresses. This new way of interacting with trees is not appropriate for every analysis, yet under certain circumstances it can be extremely useful.

Remember that a foundational use of decision trees is to partition and segment the host data set according to particular features that can be shown to distinguish between different target levels (in categorical targets) or different averages values (in targets with continuous values). Once one or more splits of the tree have been produced, a picture of the distribution of values by segments formed by the splits on the branches of the decision tree begins to emerge. Once a given segment has been identified (where the identification has been made based on the values of the branch partitions), it is sometimes useful to explore variations among other fields of data within one of the descendent nodes that correspond to the segment that you want to investigate further. This is where the capability to switch targets comes in.

With multiple targets it is possible to begin the decision tree with one target and then, at a certain point, switch the focus of the decision tree node content to reflect a different target. As an example to illustrate this technique assume that we're investigating the variations of weight in a given population. Let's say that we observe a higher average weight among male members of the population (compared to females). And, let's assume that within the male group of the sample, we see variations of weight according to the height of the male members. Let's assume we find a node with higher than average weights … and for the purposes of our example here, let's assume these comprise the tall/male segment of our population.

This example scenario is illustrated in Figure 3.28.

Figure 3.28: Illustration of Target-Switching Process

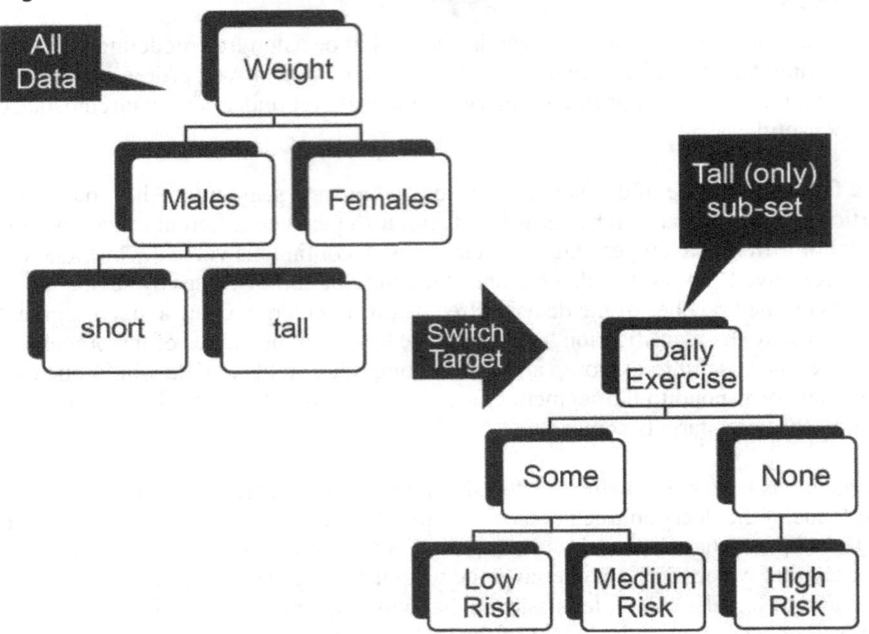

Now that we have identified this particular segment of the population, let's assume that we want to do a further in-depth study of the attributes of this segment. For example, we might want to explore variations in fitness or diet within this higher-than-average weight sub-population. We can easily do this with the "Switch targets" functionality of the SAS Enterprise Miner decision tree. In this case, we would simply define "Daily Exercise" as the new target (assume that we had recorded the daily exercise for the subjects in the data set).

To illustrate this technique with real data, let's use a live data example based on one of the demonstration data sets that are included with the SAS Enterprise Miner installation: the Home Equity data set.

Example of Multiple Target Selection Using the Home Equity Demonstration Data

Start or launch SAS Enterprise Miner (typically invoked from the Start → Programs → SAS menu selection).

Once SAS Enterprise Miner is launched, define a New Project. For purposes of our example, we define a project called "Multiple Targets." In this example, we will also use one of the pre-supplied analysis databases: Home Equity (this is usually found in the SAMPSIO directory as HMEQ).

If you right-click the **Home Equity** data source (shown under the Data Sources selection in the folder tree on the left), you see a display such as Figure 3.29.

Figure 3.29: Home Equity Data Source Attributes

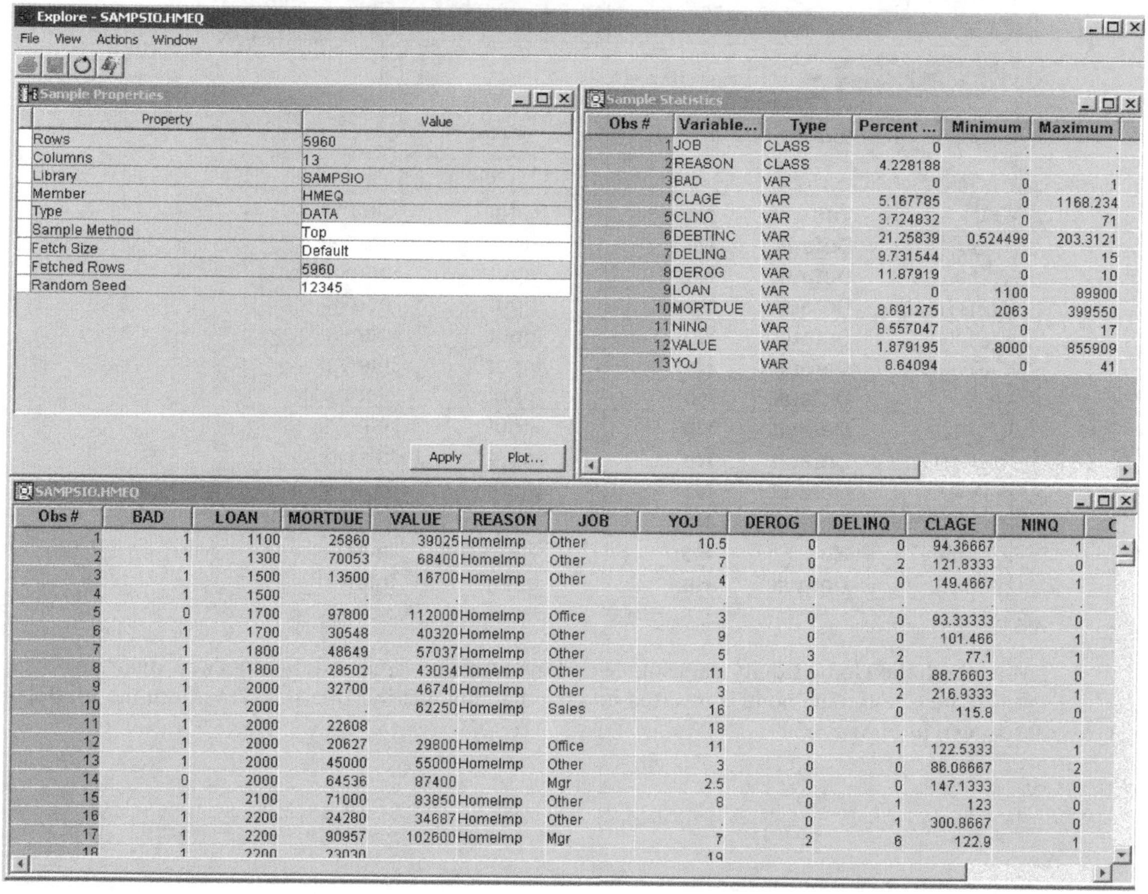

Explore - SAMPSIO.HMEQ

File View Actions Window

Sample Properties

Property	Value
Rows	5960
Columns	13
Library	SAMPSIO
Member	HMEQ
Type	DATA
Sample Method	Top
Fetch Size	Default
Fetched Rows	5960
Random Seed	12345

Apply Plot...

Sample Statistics

Obs #	Variable...	Type	Percent ...	Minimum	Maximum
1	JOB	CLASS	0	.	.
2	REASON	CLASS	4.228188	.	.
3	BAD	VAR	0	0	1
4	CLAGE	VAR	5.167785	0	1168.234
5	CLNO	VAR	3.724832	0	71
6	DEBTINC	VAR	21.25839	0.524499	203.3121
7	DELINQ	VAR	9.731544	0	15
8	DEROG	VAR	11.87919	0	10
9	LOAN	VAR	0	1100	89900
10	MORTDUE	VAR	8.691275	2063	399550
11	NINQ	VAR	8.557047	0	17
12	VALUE	VAR	1.879195	8000	855909
13	YOJ	VAR	8.64094	0	41

SAMPSIO.HMEQ

Obs #	BAD	LOAN	MORTDUE	VALUE	REASON	JOB	YOJ	DEROG	DELINQ	CLAGE	NINQ	C
1	1	1100	25860	39025	HomeImp	Other	10.5	0	0	94.36667	1	
2	1	1300	70053	68400	HomeImp	Other	7	0	2	121.8333	0	
3	1	1500	13500	16700	HomeImp	Other	4	0	0	149.4667	1	
4	1	1500	.	.				.	.	.	.	
5	0	1700	97800	112000	HomeImp	Office	3	0	0	93.33333	0	
6	1	1700	30548	40320	HomeImp	Other	9	0	0	101.466	1	
7	1	1800	48649	57037	HomeImp	Other	5	3	2	77.1	1	
8	1	1800	28502	43034	HomeImp	Other	11	0	0	88.76603	0	
9	1	2000	32700	46740	HomeImp	Other	3	0	2	216.9333	1	
10	1	2000	.	62250	HomeImp	Sales	16	0	0	115.8	0	
11	1	2000	22608	.			18	.	.	.	.	
12	1	2000	20627	29800	HomeImp	Office	11	0	1	122.5333	1	
13	1	2000	45000	55000	HomeImp	Other	3	0	0	86.06667	2	
14	0	2000	64536	87400		Mgr	2.5	0	0	147.1333	0	
15	1	2100	71000	83850	HomeImp	Other	8	0	1	123	0	
16	1	2200	24280	34687	HomeImp	Other	.	0	1	300.8667	0	
17	1	2200	90957	102600	HomeImp	Mgr	7	2	6	122.9	1	
18	1	2200	23030				19					

The main project interface has an **Edit Variables** function that will enable you to define an additional field as a target (in addition to the normal one target project). Accordingly, ensure that both BAD and VALUE (field number 3 and 12) are selected as targets. This will set up the following project steps to support the construction of a decision tree that uses multiple targets (in this case Bad and Value). It is important to set up these two targets ahead of time because when the decision tree starts to run, all the internal data structures that are built are based on the data properties of the targets and inputs. Identify the targets ahead of time to ensure that all the necessary input-target components are available to support target switching once the process flow construction begins.

The target definitions are shown in Figure 3.30.

Figure 3.30: Target Definitions for Multiple Targets

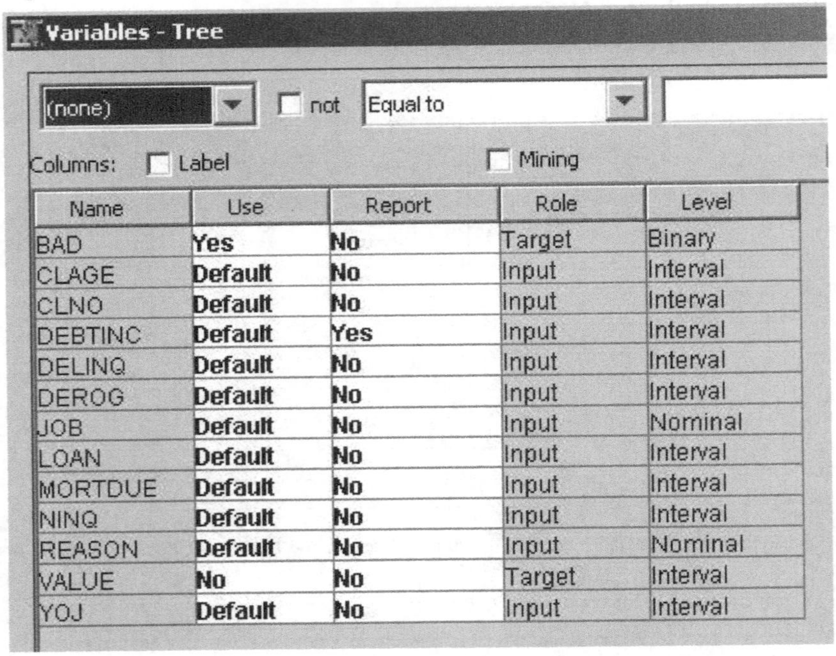

Variables - Tree

(none) | not | Equal to

Columns: Label | Mining

Name	Use	Report	Role	Level
BAD	Yes	No	Target	Binary
CLAGE	Default	No	Input	Interval
CLNO	Default	No	Input	Interval
DEBTINC	Default	Yes	Input	Interval
DELINQ	Default	No	Input	Interval
DEROG	Default	No	Input	Interval
JOB	Default	No	Input	Nominal
LOAN	Default	No	Input	Interval
MORTDUE	Default	No	Input	Interval
NINQ	Default	No	Input	Interval
REASON	Default	No	Input	Nominal
VALUE	No	No	Target	Interval
YOJ	Default	No	Input	Interval

Move both the Home Equity data source and the Decision Tree nodes onto a new diagram. Once the Home Equity data node is inserted into the diagram, get the Decision Tree from the **Model** tab, as shown in Figure 3.31.

Figure 3.31: Opening the Decision Tree Node on the Diagram

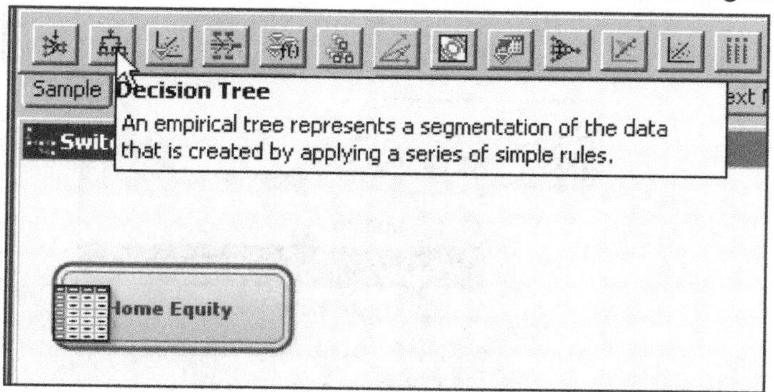

Connect the data node with the decision tree node. Click the **Decision Tree** node to explore the decision tree node run properties.

Figure 3.32: The Multiple Targets Attribute

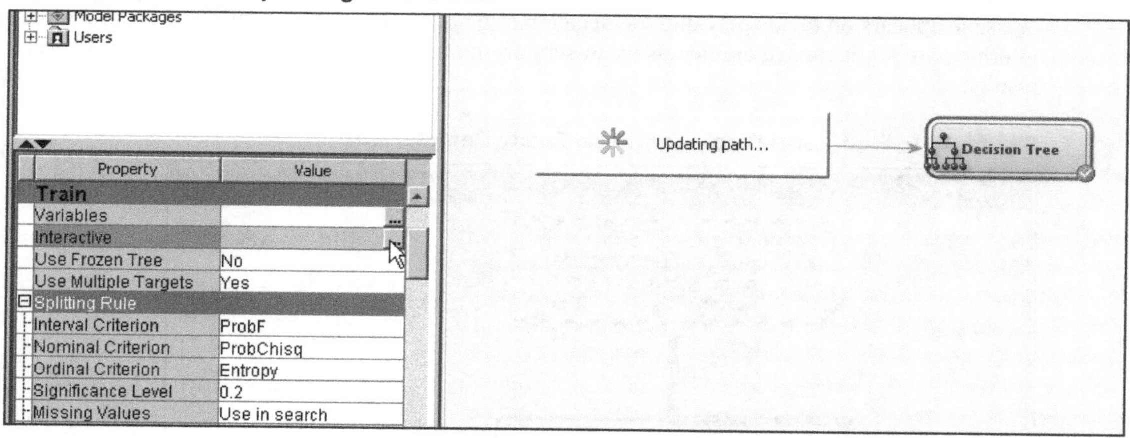

As shown in Figure 3.32, ensure that the **Use Multiple Targets** property is selected as "Yes" under the "Property" column of the display. This step takes advantage of the multiple target definition that you set earlier in the Edit Variables dialog box. When you set multiple targets in the Edit Variables dialog box and follow this up by setting the **Use Multiple Targets** property, you ensure that the underlying computer mechanisms can build the decision tree results structure with the capacity to support multiple targets.

Figure 3.33: Growing the Decision Tree

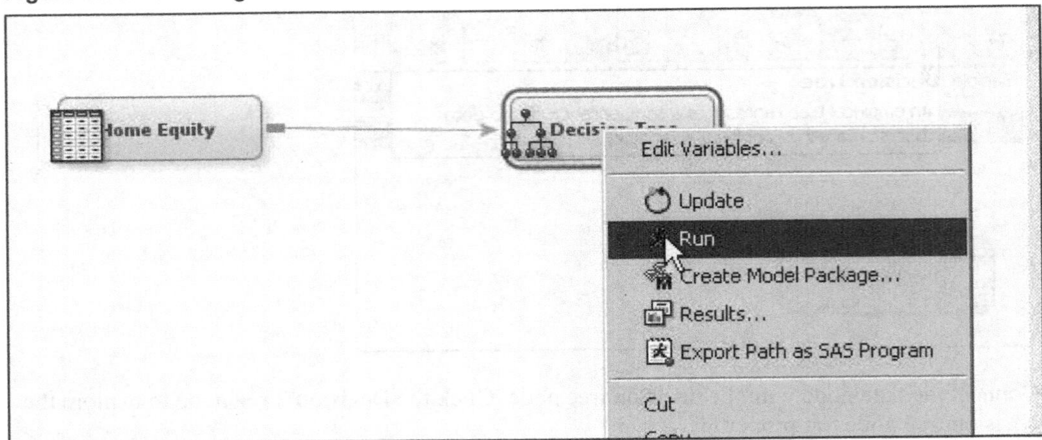

Right-click the **Decision Tree** node and, as shown in Figure 3.33, select Run. An "updating path" message appears on the display and the decision tree node calculates the various data structures and intermediate products that are necessary to support the calculation and display of multiple target results.

Figure 3.34: First-Level Display of Home Equity Default Data

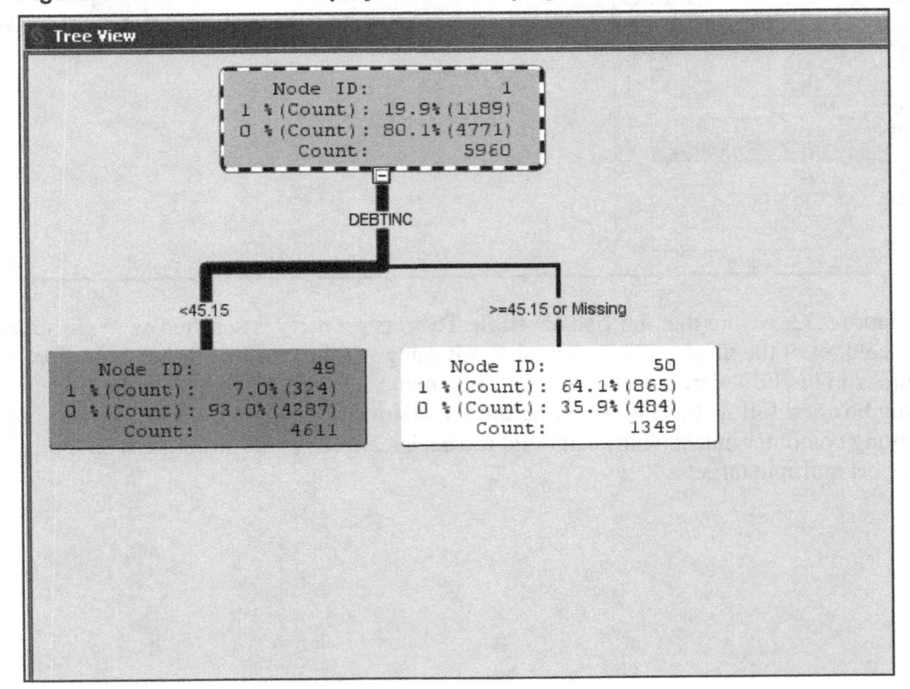

The decision tree runs in default mode and automatically calculates a decision tree structure consisting of potential splits of the root node. The attributes of the various inputs are also calculated–including such attributes as potential splitting points on continuous valued inputs–and a data structure is created that will facilitate further computations.

At the end of this calculation step, as shown Figure 3.34, the top ranked partition of the root node is presented for viewing. In this case, you see that if the debt to income ratio (DEBINC) is 45% or greater, then the percentage of defaults increases from approximately 20% in the root node to over 60% in the high debt-to-income ratio node. This is a 3-times increase in risk, or a lift of 300%.

Figure 3.35: Selecting Multiple Targets in the Interactive Mode

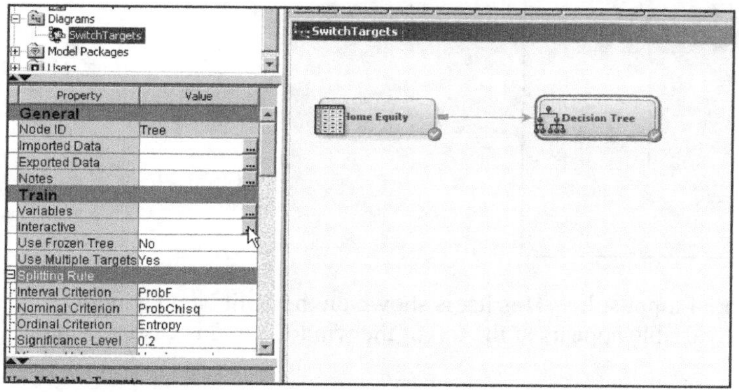

You can also run the decision tree in interactive mode by selecting **Interactive** under the **Train** menu on the left (shown in Figure 3.35). When the interactive display is presented, you can right-click on the root node and select **Split Node**.

Figure 3.36: Split Node Display

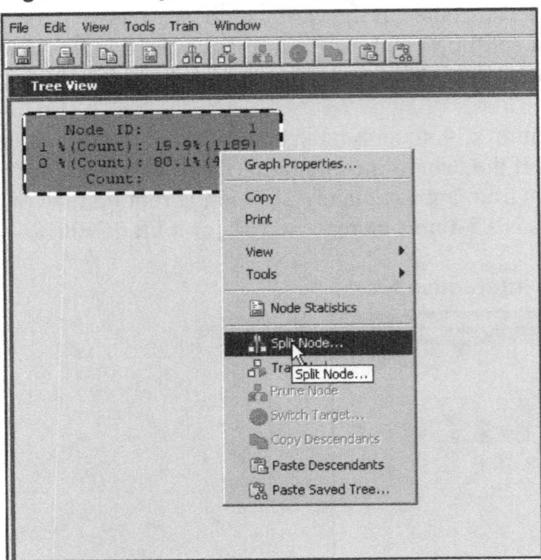

This presents a listing of potential input splits. This list is shown on the Split Node window in Figure 3.37. The current target variable appears at the top of the window; in this case, the target is BAD.

Figure 3.37: Available Views Under Split Node Window

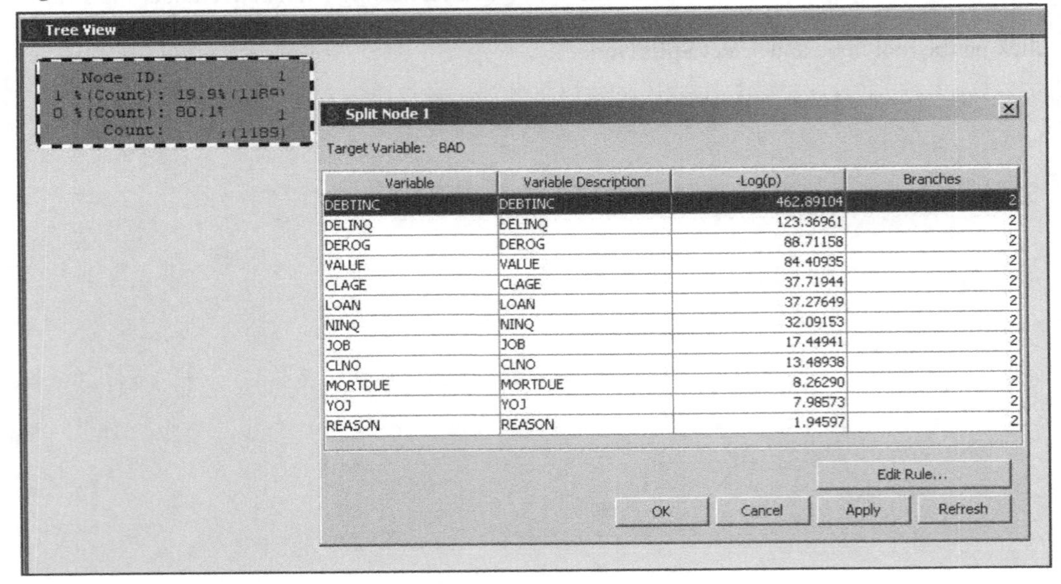

Variable	Variable Description	-Log(p)	Branches
DEBTINC	DEBTINC	462.89104	2
DELINQ	DELINQ	123.36961	2
DEROG	DEROG	88.71158	2
VALUE	VALUE	84.40935	2
CLAGE	CLAGE	37.71944	2
LOAN	LOAN	37.27649	2
NINQ	NINQ	32.09153	2
JOB	JOB	17.44941	2
CLNO	CLNO	13.48938	2
MORTDUE	MORTDUE	8.26290	2
YOJ	YOJ	7.98573	2
REASON	REASON	1.94597	2

As shown in Figure 3.38, once the input split is selected, a number of plots and statistics are available under the **View** menu. For example, select **Leaf Statistics Bar Chart**.

Figure 3.38: Decision Tree Display

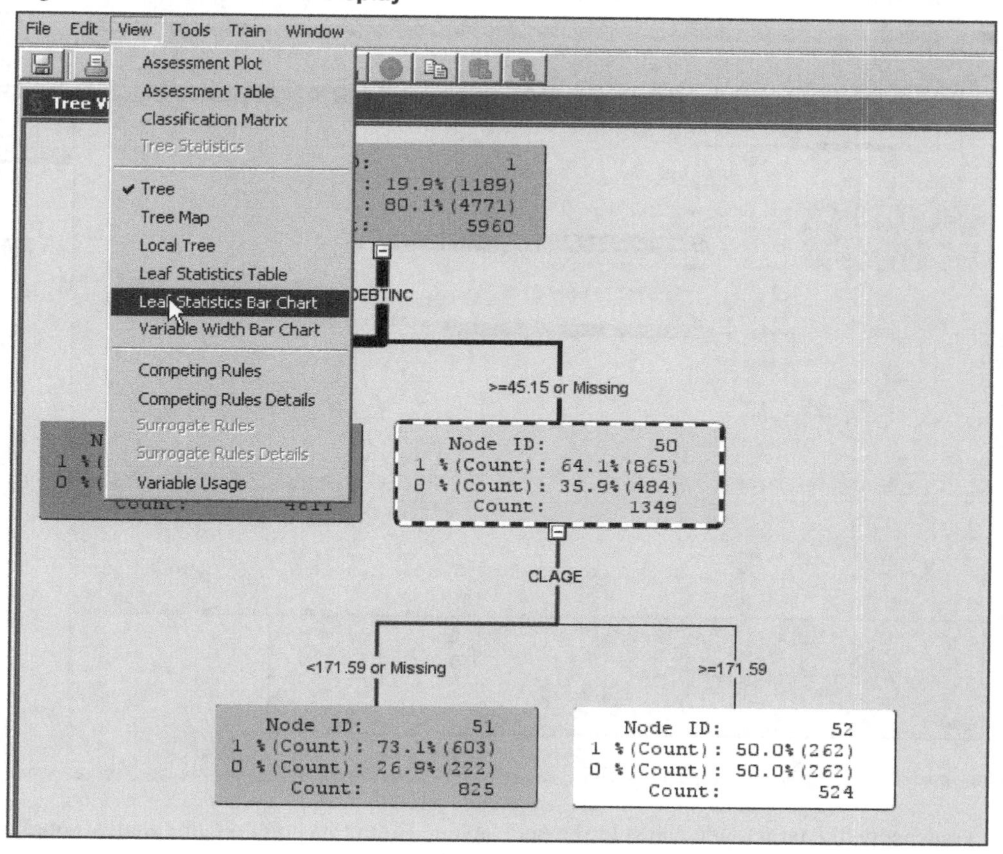

Figure 3.39: Selecting the Switch Target Process

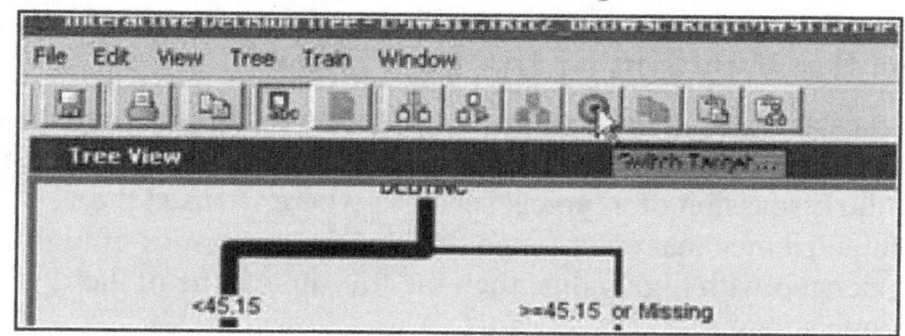

As shown in Figure 3.39, there is a target icon on the menu bar of the interactive menu. This allows you to change focus and switch targets.

Figure 3.40 shows how to change the target to VALUE. This will affect the descendent nodes that are grown from the left-hand node that was formed by splitting the root node.

Figure 3.40: View of Mid-Tree Target Switch

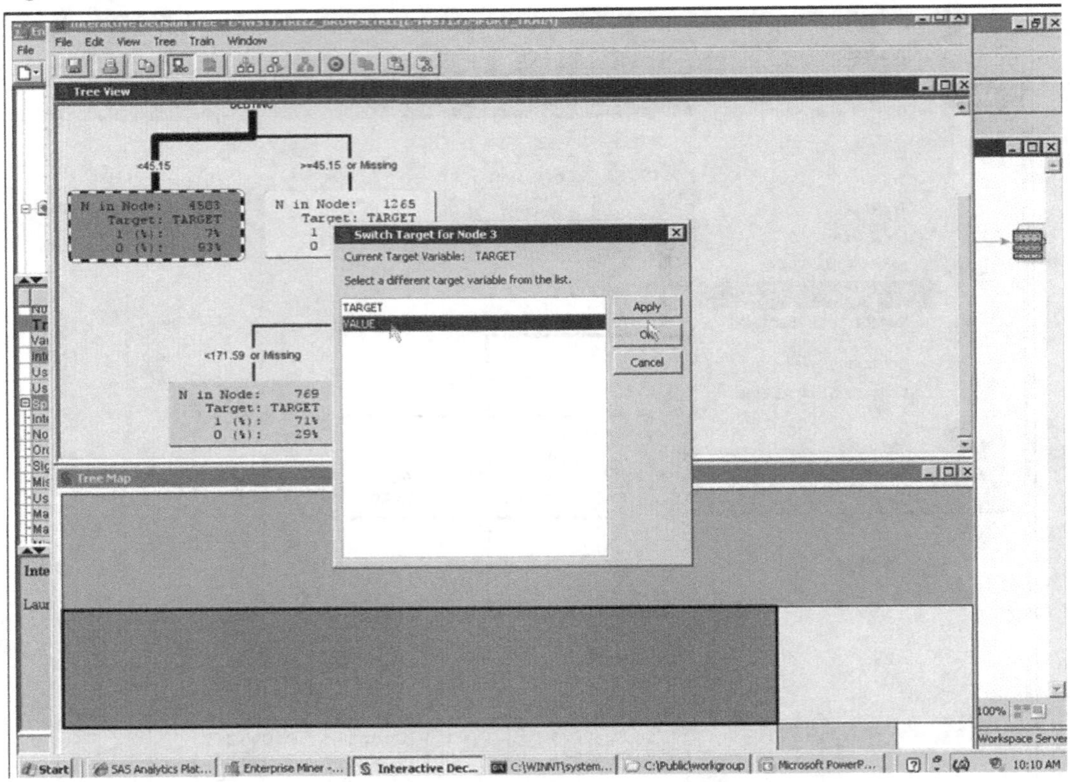

After you change the target, you can split the node again. From this point on, our analysis goal begins to change from predictive modeling to strategy development.

Synergy, Functionality, and the Wisdom of the End User

We had been working with decision trees as a commercial offering for a little over ten years when this functionality was first suggested to us by a user who was a complete novice in using decision trees. The suggestion evolved during the discussion of a typical scenario where we used the decision tree to carve out the important attributes that determined the key characteristics of high-tenure customers. If high tenure is associated with high value, then the life-time value of the customer represents a significant revenue stream.

We had identified a node in the tree where tenure was over five times the average across the entire data set. The idea that jumped into the new user's mind at that point was to display the average purchase amount that the customers in this node made, and then use the decision tree again to tease out and differentiate the high-value versus low-value customers. As developers, we had done this kind of analysis many times through the "normal" process of exiting the decision tree software to go back out to the computer's file system. Once in the operating system, we could retrieve the source data set, and, by sub-setting the data based on the original decision tree split rules, we could create an analytic sub-set that we could then use to create a separate analysis.

This new suggestion was at odds with our normal practice at the time but was also obviously a great time saver for the end user. Years later, the core functionality of decision trees moved more and more into the hands of business users. These users were not well versed in data manipulation methods, so they could only go back to the original data for purposes of sub-setting with great difficulty. At this point, it turned out that the ability to switch targets at any point in the decision tree growth was a highly prized capability that can now be considered a mandatory feature of decision trees.

---

[1] The 95% level of statistical confidence asserts that the difference in the two distributions is so sufficiently large (or abnormal) that you would see this effect only about 5% of the time by chance alone. Because the likelihood of seeing it by chance is small (i.e., 5%), the statistical conclusion is to assert that the relationship is significant.

[2] The X^2 test was developed for row X column tables. It is appropriate for a decision tree because a partition on a decision tree is, in fact, a row X column table (where the values of the target form the rows and the values of the partitions form the columns). This similarity between decision trees and tables is reflected in the use of decision trees for multidimensional cube analysis. In fact, a decision tree can be viewed as a multidimensional cube.

Chapter 4: Business Intelligence and Decision Trees

Introduction

Business Intelligence (BI) applications have been one of the fastest growing applications in the early years of the twenty-first century. BI applications provide a set of tools and techniques to enable the storage, retrieval, manipulation, and display of data to domain experts and business analysts in a form that facilitates business and policy decision-making. BI applications are part of a maturity evolution in the computer industry that has enabled the use of computers to move from operational applications that can be used to run the business, to analytical applications that can be used to drive the business and steer its direction.

A key feature of BI is the deployment of a wide range of readily available reporting capabilities. Deployment is achieved without obvious intervention of specialized IT staff. The consumers of the information contained in the reports have a much higher degree of access than was possible before the development of BI. As a result, BI has become a pervasive business tool and approach since its origin in the mid-1990s.

Key drivers of BI use include:

- the development of data warehousing concepts and techniques (to access data and to combine multiple data sources to form a view of data that can be consumed by BI computer agents)
- the development of data and dimensional storage and retrieval capabilities that have been adapted to serve in the BI reporting engines
- the evolution of a wide variety of data viewing techniques, including the production of reports, spreadsheets, business graphics, and Web deployment environments

More recently, BI has given rise to the identification of business analytics. The term business analytics explicitly recognizes that there is so much data available and that there are so many factors involved in business processes and business decision-making that analytical approaches and techniques are a necessary underpinning for BI to perform effective data summarization and trend identification. In the beginning of BI, its analytic nature and the real-time deployment of results to the user led to a description of the area as OLAP (Online Analytical Processing). The term OLAP, which is still used, has given way to multidimensional cubes, which provides a broader description of the area.

Traditional BI tools enable an analyst or decision-maker to display multiple views of multiple items of interest. For example, BI reports and spreadsheets show sales by region, sales by time, or sales by product line. With the increasing incorporation of business analytics within the BI framework, there are more methods and mechanisms that use business analytics in the identification, explanation, and dissemination of BI results.

For example, many views can be derived from the following display:

Credit-Worthiness	Mortgage Due	Home Value	Length of Residence	Age
1	25860	39025	10.5	64
1	70053	68400	7	21
1	13500	16700	4	49
0	97800	112000	3	60
1	30548	40320	9	31
1	48649	57037	5	47
1	28502	43034	11	38
1	32700	46740	3	42
1	20627	29800	11	28
1	45000	55000	3	36
0	64536	87400	2.5	47
1	71000	83850	8	40
1	24280	34687		31
1	90957	102600	7	23

1	28192	40150	4.5	54
0	102370	120953	2	45
1	37626	46200	3	62
1	50000	73395	5	
1	28000	40800	12	67
1	17180			56
1	34863	47471	12	34

This display can be viewed as cube slices, as shown in Figure 4.1.

Figure 4.1: Illustration of Cube Slices Defined for BI Display

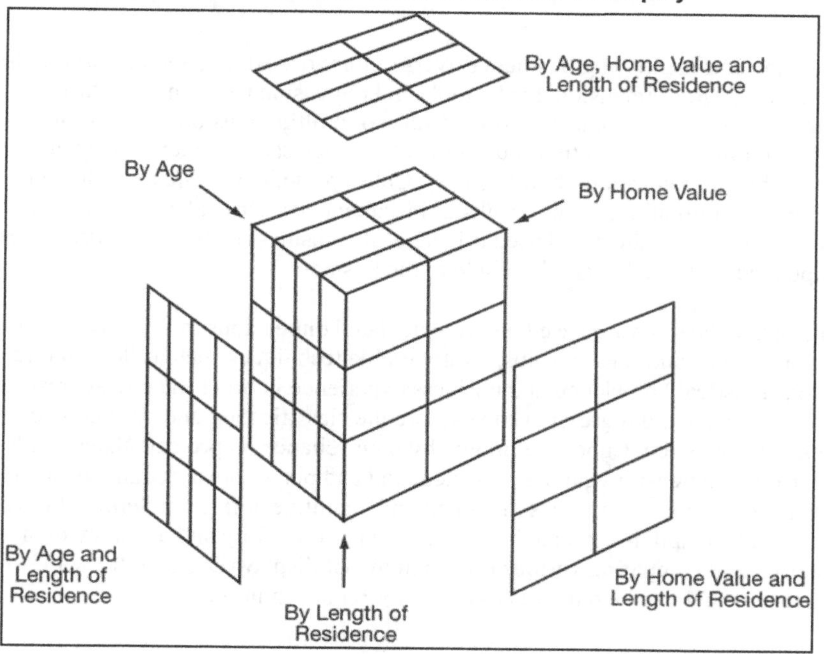

When viewed this way, you can see that the cube concept is dominant in BI. Many advances have been made in both the front end (data presentation) and back end (data warehousing, data integration, and data manipulation) aspects of BI to handle the range of records, fields, field values, and associated observations, variables, and variable values that are richly reflected in the enterprise data store. BI products are timelier, more accessible, and more flexible than ever before.

As the state of the art continues to evolve, there has been a corresponding demand in the area of business analytics. Improvements in techniques in the areas required to perform prospective tasks in BI are needed:

- Provide multi-factor versus single-factor approaches and displays. Until recently, it was common to show 2-way diagrams (sales by region) or 3-way diagrams (sales by region by product line). The ability to show multi-factor hot spots and soft spots has been difficult to produce (for example, produce a display that shows that although sales in the west for new products are low, this is not the case for special segments of the population).
- Provide forward-looking, predictive, or what-if approaches versus historical, snapshot-of-the-past approaches.

The tasks of understanding and identifying business drivers require classifying and confirming trends and relationships in the data. Multidimensional cube tools and regression perform these tasks. Multidimensional cubes form the underpinnings of BI. Cubes or multidimensional cube tools enable business users to look at multiple views of their business data as they seek a better understanding of the trends and relationships that are relevant to their business. Cubes provide pre-calculated and pre-summarized dimensions of information, which results in instantaneous retrieval and examination. The ultimate goal is to better understand the data-based drivers of the business so that these drivers can be anticipated and manipulated in ways that are favorable to the business.

Whereas multidimensional cube tools are based on pre-calculated dimensions to improve a user's judgment when assessing trends and relationships, multivariable techniques such as decision trees and regression are based on statistical knowledge and business experience in order to generate results on the fly. Multidimensional cube tools and regression can explore the classification and predictive power of multiple fields of data in a data store. Cubes are limited by their reliance on pre-calculated fields; simply put, not all relevant business dimensions can be pre-calculated and pre-summarized in a business analysis. There are significant limits to a user's judgment and cognitive abilities in terms of the number of quantities that can be judged and manipulated, as well as in the reliability and accuracy of size estimates, when exploring and comparing various effects in the BI display. Decision trees are well adapted to producing results that can be rendered as cubes for reporting purposes.

A Decision Tree Approach to Cube Construction

Like BI tools, decision trees perform the tasks of trend and pattern identification. Decision trees are built using a methodology that explicitly addresses the need to identify the relationships between the factors that combine to provide a complete view of the area being examined. Decision trees are designed to search for a wider range of relationships than multidimensional cubes or standard regression methods. Because decision trees drill down to the record level in data, they enable multidimensional business reports that identify trends and patterns that might be missed in typical, multidimensional cube and regression analyses that rely on aggregate data.

The BI report is based on displaying facts (usually from the fact table of the underlying data warehouse) along preselected dimensions and sub-dimensions. BI reports are defined ahead of time to reflect commonly used business reporting dimensions. Decision trees also display facts (usually from the distribution of a target field) along dimensions and sub-dimensions, which are formed by the branches of the decision tree. The major difference between the dimensions of a decision tree and the dimensions of a BI approach is that the dimensions of a decision tree are formed on an ad hoc basis, either automatically or through user interaction. Decision tree displays result from compressing or collapsing dimensional values on the display. Values of the dimensions and sub-dimensions are collapsed to show similarities and differences among the dimensional values that are not highlighted in BI displays.

Decision trees evolved as data analysis tools in both applied and academic settings. The earliest use of decision trees was in a marketing research analysis that involved an audience. Other early uses were developed to assist the identification of relationships in data to support sociological and economic research at the Survey Research Center at the University of Michigan. Decision trees have had a strong business analysis orientation from the early days of their conception.

Decision trees began as a method of finding tables within tables or relationships within relationships. In this respect, they are like multidimensional cube tools in that they both look at various dimensions of data and, within a dimension, they both look within sub-dimensions. As decision trees matured, the goals extended to handle continuous and categorical table cell entries and multi-way branches. Statistical tests and validation approaches were developed to assure the integrity of the decision tree. Decision trees use data search and summarization algorithms and verification and validation mechanisms that distinguish them from multidimensional cube tools.

Multidimensional Cubes and Decision Trees Compared: A Small Business Example

Assume you have a lemonade stand and are selling lemonade by the glass or by the jug. You have a database of sales transactions. You even have a number of fields of information where you collect additional data each time you make a sale.

A sales transaction record might appear as follows:

Sale Item	Quantity	Price	Discount	Time of Day	Sales Agent	Customer Identifier

You use the database to calculate sales commissions and to keep track of inventory to reorder supplies. Today, with the growth of analytical systems, you could use this data to try to understand your customers and your sales patterns, to create sales campaigns, and to drive new product development. You could look for types of sales that maximize the profit from your sales effort (return on investment, or ROI).

A typical multidimensional cube analysis starts with historical business queries and reports. It attempts to identify dimensions in the data that elaborate the business model to create views of the contextual effects. These views lead to a better understanding of the relevant business issues. The multidimensional cube analysis proceeds intuitively—it follows hunches to look for relationships that can be used to better understand or predict events.

Consider your database of lemonade sales to determine sales trends. A multidimensional cube analysis might look at sales by time of day, as shown in Figure 4.2.

Figure 4.2: Illustration of a Multidimensional Cube by Time of Day

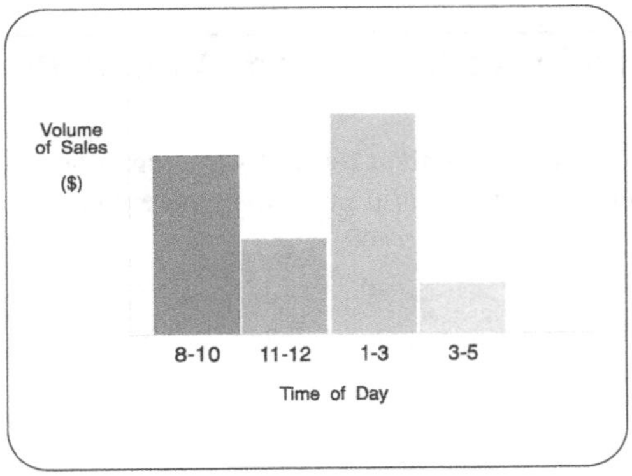

Volume of Sales ($)

8-10 11-12 1-3 3-5

Time of Day

Or, the analysis might characterize customer by customer type.

Figure 4.3: Illustration of Age Characteristics of Customers

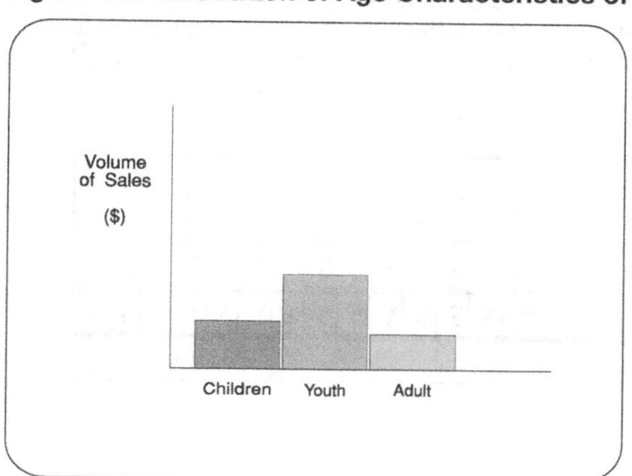

Or, the analysis might break down sales by volume.

Figure 4.4: Illustration of Sales by Volume

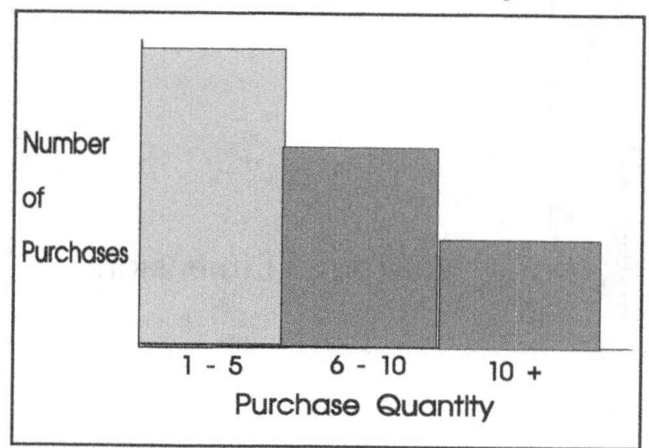

The analysis might even look at sales by geographic area served (customer origin).

Figure 4.5: Illustration of Customer Origin

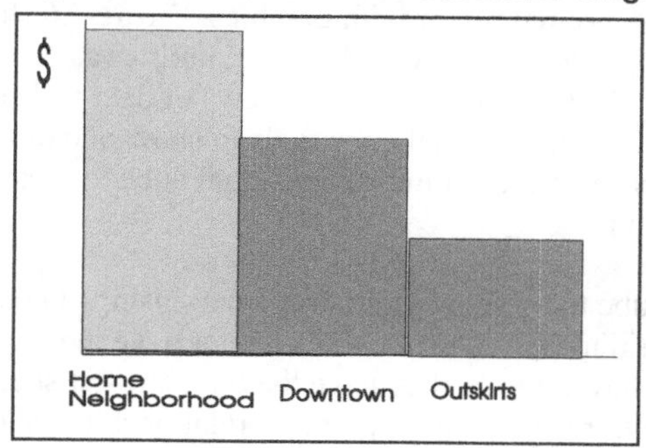

Behind the scenes, a typical multidimensional cube analysis is based on a cube that has dimensions of analysis that have been determined as being important characterizations of the business data. Figure 4.6 contains two simple examples—one in two-dimensional form and one in three-dimensional form.

Figure 4.6: Multidimensional Cube Analysis

Customer	$								$							
Quantity	$				$				$				$			
Location	$		$		$		$		$		$		$		$	
Time of day	$	$	$	$	$	$	$	$	$	$	$	$	$	$	$	$

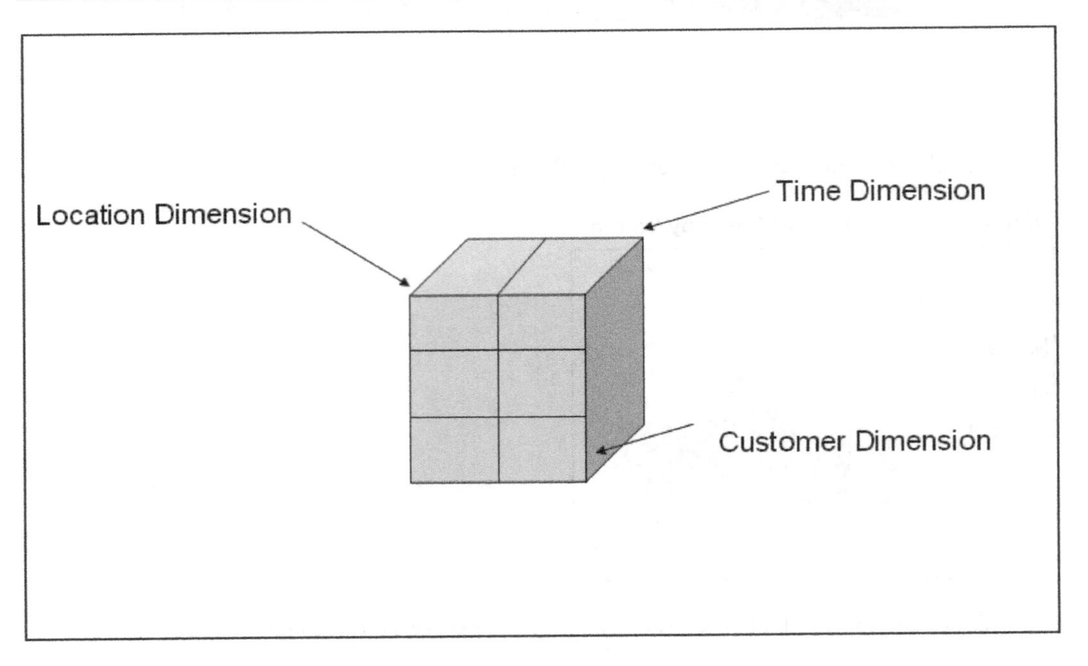

In these examples, the business user is using the cube definitions and associated data tables to present views of the data. Views could be a screen display or a printed report with graphics. Figure 4.6 shows 2-way relationships: one field on one axis, and one field on the other axis. However, multi-way relationships are possible and desirable. The ability to drill down into various views of data and show multi-way relationships within a dimension enables multidimensional cube tools to show one or more relationships in the context of another relationship; for example, a multidimensional cube tool can show the discount rate for a product for an enterprise division.

Although both decision trees and multidimensional cube tools show multi-way relationships in context, there are important differences between the two; multidimensional cube tools do not have the same relationship-searching capabilities that decision trees have, nor do they have the onboard statistics or validation facilities that decision trees have. A decision tree looks through more relationships than a multidimensional cube looks through; furthermore, a decision tree verifies and validates relationships as being statistically or numerically sound.

Multidimensional cube views are designed to support quick viewing and decision-making by the business user. As a result, the cubes are built to optimize the user's time. Much care and effort are required in constructing the underlying multidimensional cube database and in precomputing the contents of the views, sub-views, and drill-downs that the business user is likely to review. This means that the business user is not able to point and click through various alternative analysis scenarios within the multidimensional cube environment, as can be done with decision trees.

Because the dimensions of a cube analysis are created from a pre-existing warehouse and because the associated reports are often pre-calculated, multidimensional cube reports tend to be more structured and rigid than decision tree reports. For example, typical state aggregations can be drawn along regional lines—East, West, South West, and so on. These dimensions can be fixed by business rules and business policies. Nevertheless, once the dimensions are established and set up either in the data warehouse or as reporting dimensions in the multidimensional cube reporting application, they cannot easily be changed, nor can they be recombined based on a relationship in the data (for example, combine all high-margin states).

Decision trees permit the recalculation of dimensional groupings on the fly. They support dimensional groupings that are based on the properties of data and the relationships between data elements, rather than on business rules or business policies. For example, the following decision tree shows how trees can be used to form the framework of a cube.

Figure 4.7: Illustration of a Decision Tree as Raw Material for Cube Construction

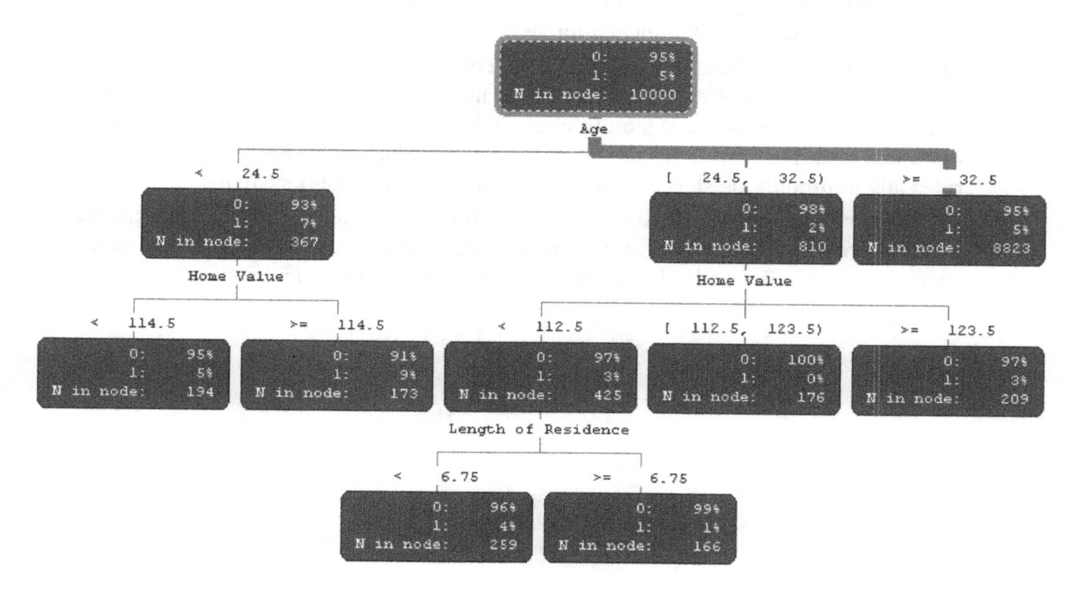

Decision trees provide utilities to tell you which variables are important and which interactions (i.e., which cube faces and cube combinations) should be presented. From Figure 4.7, you can see many useful features for the display of cubes.

The decision tree identifies empirical-based ranges of values to collapse (or group) within display fields, such as age. Normally, these ranges are predetermined or calculated mechanically. The decision tree identifies whether sub-dimensions are significant (from a statistical point of view, and also from a business-rules perspective) and, if so, what the optimal collapsed categories for the values should be.

In Figure 4.7, you can see that at the top level of the decision tree, the optimal code ranges for **Age** are < **24.5**, **24.5** to **32.5**, and >= **32.5**. The structure of the decision tree suggests that two sub-dimensions, based on **Home Value**, are appropriate in the lower **Age** range. **Length of Residence** is an appropriate sub-dimension in the low-end value of the middle **Age** range.

Multidimensional Cubes and Decision Trees: A Side-By-Side Comparison

Multidimensional cubes and decision trees can be compared and contrasted as follows:

- While multidimensional cubes have pre-built data dimensions, the dimensions of decision trees are dynamically collapsed to highlight similarities and differences among and between nodes that are being formed by the decision trees. These nodes are equivalent to classes in the cells of the tables.
- In decision trees, the dimensions and groupings can be determined by business rules, as BI approaches are, but they are more commonly determined by the strength of the association or prediction, which is based on numerical methods or statistical approaches. Typically, the dimensions and groupings are formed through a dynamic that combines business rules with numerical methods and statistical approaches.
- Multidimensional cubes are almost always retrospective; that is, they show what has already happened, based on the data. Decision trees are retrospective, prospective, and predictive. Like multidimensional cubes, decision trees show how the data is arranged in-line with the historical past. Decision trees can be used to extrapolate and infer future events. Underlying rules are commonly used as the basis for predictive and expert systems.
- Decision trees are more effective at handling missing values. Missing values are handled by using surrogate or stand-in values or by treating missing values as a different code that is grouped with similar codes in the reporting dimension of the decision tree.

Table 4.1: Comparison of Multidimensional Cubes and Decision Trees

Multidimensional Cube	Decision Tree
Shows tabular views of data as tables with relatively fixed dimensions; dimensions are determined primarily on the basis of business rules	Shows tabular views of data within relevant dimensions as determined by computational algorithms and business rules
Has database that is pre-built to support anticipated queries	Has database that is pre-built to support numerous unanticipated queries
Provides quick view retrieval	Has lengthy retrieval
Tends to limit number of cross-views or relevant factors	Has few limitations on the relevant factors
Makes it difficult, almost impossible to identify novel results	Emphasizes novel results and the identification of important versus unimportant contributions

Decision tree results can be made to look very much like multidimensional cube results; the branches of a decision tree are just simple n-way tables that show the relationship between the attributes of the field that is used to form the branch and the values of the target in that node or leaf. Like multidimensional cube tools, decision trees can display both categorical and continuous n-way relationships in any node or leaf. Like multidimensional cube tools, a leaf is presented in the context of higher-level dimensions. In a decision tree, these higher-level dimensions are the higher-level branches. In a multidimensional cube, the higher-level dimensions represent the drill-down sequence that was followed in order to be at that face in the cube.

Both decision trees and multidimensional cube tools provide a drill-through capability (i.e., the ability to display and analyze detailed information that belongs to the individual records that characterize the relationships that are displayed in any single table, leaf, or face of the cube). And, just as decision trees can be represented as multidimensional cubes and associated displays, multidimensional cubes can be represented as decision trees (although typically, they are not).

Both multidimensional cubes and decision trees provide the means to apply all relevant dimensions when identifying key drivers that affect a target or outcome value. However, multidimensional cube dimensions are displayed and examined hierarchically, whereas decision trees present results in tree form as a network. In summary, both support a style of analysis that can lead to identifying important relationships between fields or variables that need to be considered in order to accurately and reliably describe and predict a target or outcome.

Multidimensional cubes present numerical results (such as average, standard deviation, mode, range, and count) within the cells that are formed by the dimensional categories. Decision trees can present numerical values and categorical values as the target.

The Main Difference between Decision Trees and Multidimensional Cubes

The major difference between decision trees and multidimensional cubes is the heavy concentration of statistical and search algorithms that are built into decision trees. All forms of multidimensional cube analyses depend on the creation of a view of the analysis data. The view enables the dimensions of the cube to be retrieved and assembled as the various faces of the cube are selected for analysis and display. In this respect, decision trees require less preprocessing of the data and of associated dimensions because the statistical and search algorithms have been built to identify the specific form of dimensions at the time that the decision tree is grown.

Decision trees support a looser initial definition of the dimensions of data that are included in the analysis. They support a more dynamic identification of the specific structure of the dimensional relationships through the use of the statistical and search algorithms. These algorithms assemble the dimension in real time as the decision tree is grown.

The actual statistical and search algorithms in decision trees are another major difference between decision trees and multidimensional cubes. Decision trees provide more methods to identify the strength of relationships than are provided in multidimensional cubes. Furthermore, decision trees provide more methods to guard against overfitting the data (i.e., decision trees provide methods to identify inaccurate and unreliable relationships, which is not usually provided in multidimensional cubes).

Overall, decision trees can uncover more relationships and more effects—based on unique groupings—than multidimensional cubes can. In addition, decision trees provide more options to check the efficacy of relationships that are discovered and displayed, which helps prevent overfitted relationships and the misunderstandings and unreliable predictions that are associated with them. Some multidimensional cube tools provide forecasting or predictive features, but they do not provide as many options and validation functions as decision trees provide.

Regression as a Business Tool

The use of regression techniques has long been a mainstay in scientific research and statistical process control. The pervasiveness of computers and information technology in business environments has created a situation that is ripe for uncovering uses for regression in business data analysis. Although regression is used in relatively specialized areas such as supply chain management, statistical process control, and database marketing, it has not been adopted as a business intelligence tool, even though it would address the major prescriptive requirements for BI approaches to provide forward-looking multiple-indicator results. The same could be said about more recent techniques such as neural networks. Neural networks could serve as a flexible family of nonlinear regression and discriminant analysis techniques. They are in the same class of methods as regression techniques.

Although the multidimensional cube is the most prevalent type of data analysis tool used in business settings, regression is used very often. And while the multidimensional cube is a recent data analysis tool, regression has been used in business and scientific settings for many decades.

Regression describes the relationship between two quantities in the form of an equation where one quantity—**Y**—is viewed as a function of the other quantity—**X**. This simple relationship can be shown as follows.

Figure 4.8: Illustration of a Typical Regression Result

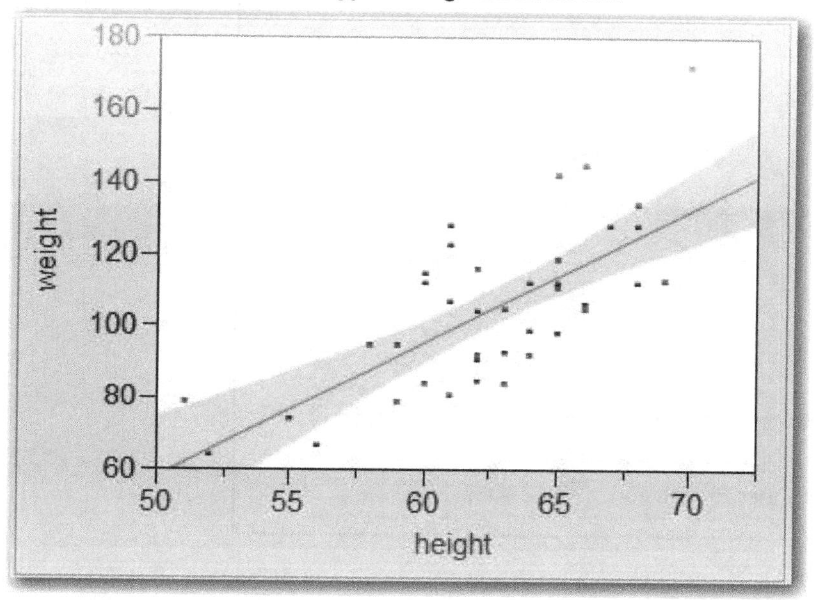

In this simple case of regression, the relationship is approximately linear. The regression method has been extended in many ways. For example, the target—**Y**—can be a function of multiple predictors. The form of the relationship has been extended so that both linear and nonlinear relationships can be included. And, in addition to numeric quantities, nonnumeric qualitative information (i.e., categorical data) can be included.

Decision Trees and Regression Compared

Decision trees and regression share a common form where target values are associated with multiple input values in order to show the form of the association and to be able to predict new target values based on new input values. Although regression and decision trees perform the same function, which is to display a relationship between a target (outcome or response) variable and one or more input variables, they take widely different approaches. Regression works by manipulating an entire matrix of information that contains all the values of all the inputs against the target and that attempts to compute an optimal form of the relationship that holds across the entire data set.

Decision trees proceed incrementally through the data. Because of this approach, a decision tree might find a local effect that is very interesting and would be missed by regression. Yet, because it is a local effect, it might be only locally significant or locally reproducible, meaning it will not replicate or generalize very well. New approaches such as boosting and bagging (which present the averaged results of many decision trees) offset this tendency.

Figure 4.9: Illustration of a Linear Trend Suggesting a Good Regression Fit

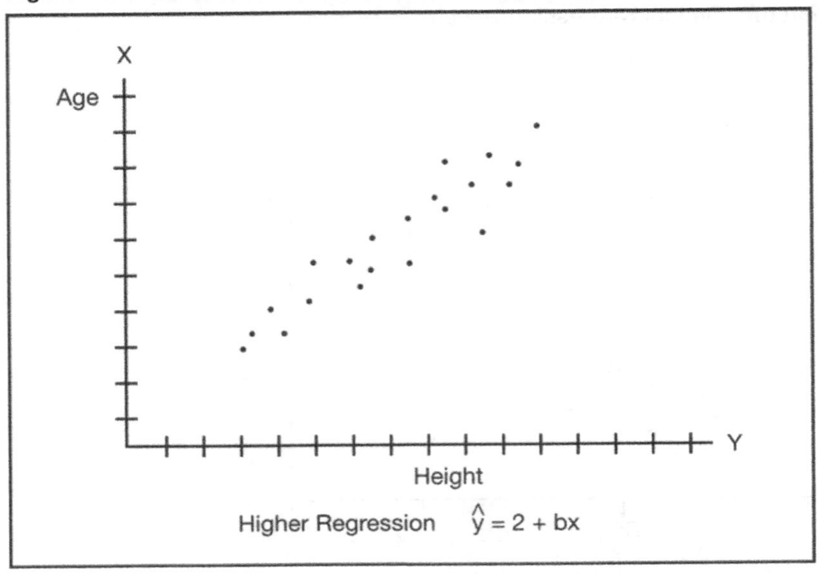

Regression is capable of presenting a linear relationship, as shown in Figure 4.9. A simple relationship can be eloquently expressed as a linear equation. Decision trees can only approximate this relationship, as shown in Figure 4.10.

Figure 4.10: Illustration of Decision Trees Displaying a Linear Trend

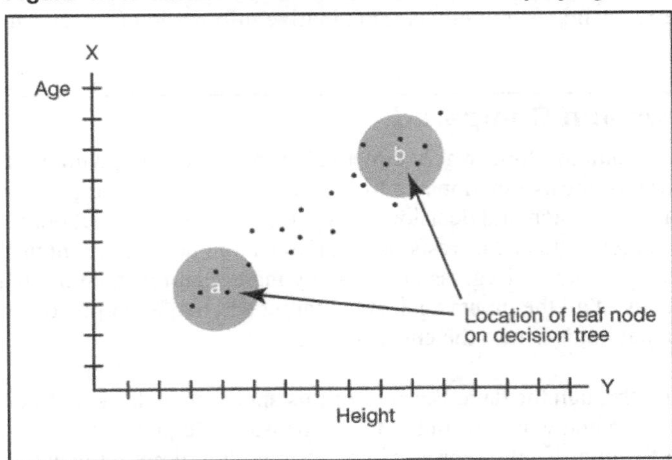

When trying to reproduce a linear relationship with decision trees, various parts of the linear relationship are fit by many components of the decision tree. This produces a staircase-type of relationship fit, as shown in Figure 4.11.

Figure 4.11: Illustration of the Staircase Effect When Multiple Decision Trees Fit a Linear Trend

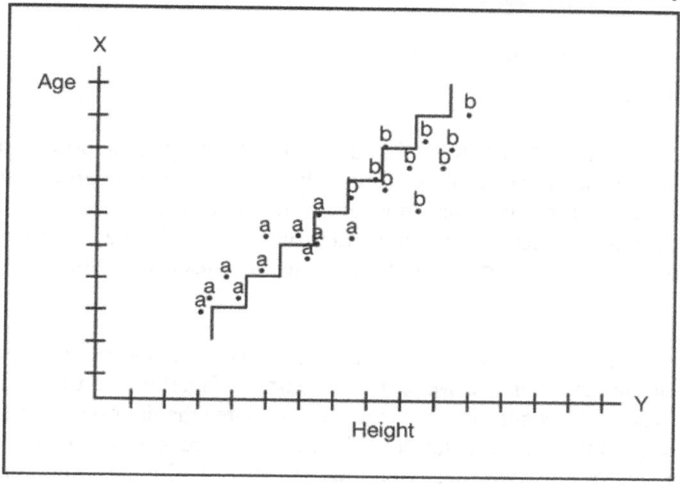

Under certain circumstances, this staircase effect can be useful (for example, when the audience has difficulty conceptualizing the regression equation). Or, this effect can be more visual when the branches on the staircase represent convenient and well-understood conceptual categories (e.g., child, preadolescent, adolescent, young adult, and so on). Often, a decision tree reveals just enough of the necessary information to be easily and intuitively interpreted, as shown in Figure 4.12.

Figure 4.12: Illustration of an Intuitive Decision Tree Displaying a Fitted Linear Trend

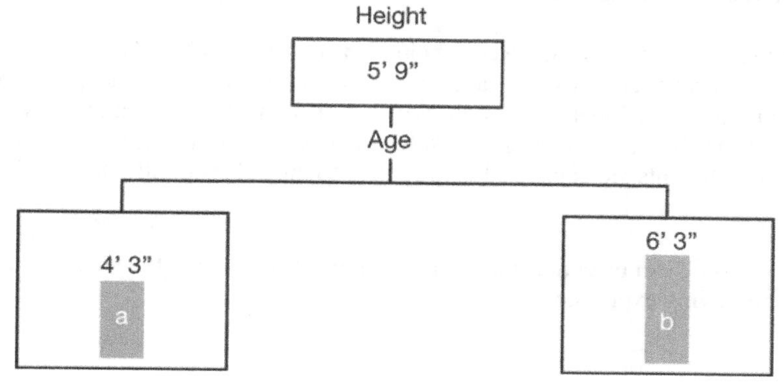

New approaches that handle many decision trees at once (e.g., boosting and bagging) offer a method of producing a smooth surface like you see with regression.

Differences between Regression and Decision Trees

The differences between regression and decision trees, as well as the indicators of the strengths and weaknesses, can be described as follows:

- local versus global search
- rules versus coefficients
- distributional and metric assumptions about the data
- description of the relationships

Local versus Global Search Decision trees attempt to find a function that can split the observations of a target (outcome or response) into subgroups that form descendent branches. These subgroups are candidates for further splitting until some stopping criterion is met. When a decision tree processes a data set, it does so in successive increments where each increment produces a leaf node that becomes a local subset where the relationship between the response and the input variables is described in a local context.

On the other hand, regression attempts to find a function that can characterize the observations of a target (outcome or response) so that the deviations between the score of the function and the actual function are minimized across all cases. The regression approach manipulates the data set and the input variables that it contains as a single canonical representation. While decision trees proceed one branch at a time to identify combined effects, regression identifies combined effects of all inputs simultaneously and then identifies individual effects. More differentiated functions are introduced into the regression equation to construct an optimal deviation-reduction function. The regression approach is global in contrast to the local segments of decision trees.

Rules versus Coefficients Decision trees proceed to segment data on an incremental basis by descending to lower branches of the decision tree. Unlike regression, decision trees look at each input separately and iteratively. They recursively choose between alternative groups or branches of input values to grow the decision tree by splitting nodes in-line with groups or branches that are identified.

By comparison, regression techniques are holistic. Regression identifies the combined and individual effects of data on the basis of matrix operations that capture and summarize the relationships between inputs and target as a single multidimensional expression. While decision tree results can be described as a series of incremental <IF> <THEN> rules, regression results are described as a series of coefficients for the model inputs. These coefficients are computed with respect to the values of all other coefficients that are in the model.

The regression equivalent of a node can be conceptualized as a slope. The value of the slope is captured by the coefficient ***bx*** in the following expression:

$$\hat{y} = 2 + bx$$

This expression reveals the upward trend in the distribution of ***x-y*** points in the regression line, as shown in Figure 4.8.

In their simplest forms, decision trees and regression equations perform different and complementary functions.

- A decision tree provides a graphical representation of the structure of the relationships in data. The decision tree identifies how the target rises and falls as the data that is associated with the target is filtered through views provided by the multidimensional breakdowns represented as leaves in the decision tree.
- A regression equation provides a clear and mathematically rigorous expression of the form of the relationships in data that is reflected by the sign and strength of the coefficients in the regression equation.

Decision trees were originally developed as a complement or alternative to linear regression. One of the earliest decision tree implementations of AID was used to detect nonlinear effects and interactions among predictors in a regression equation. Decision trees readily select sub-segment effects in a data set that might be missed by regression.

Decision trees deliver their sub-segment effects through their recursive partitioning method. This offers the advantage of more readily selecting sub-segment effects, but it does so at the cost of requiring much data to work with. As decision trees successively partition the data set, smaller sub-segments of data are created at lower levels of the branches of the decision tree. Because regression computes the combined effects of all data points through a summary operation that works with all data points simultaneously, regression does not dice the data in the same way decision trees do, and it makes more efficient use of the available data. Neural networks, like decision trees, readily select nonlinear and sub-segment effects that are contained in the data.

Distributional and Metric Assumptions About the Data A major difference between decision trees and regression is the use of categorical or nonmetric data values. This difference underscores a general difference between decision trees and regression that relates to the assumed form of the data that underlies the approaches. Because decision trees successively segment input values based on discrete, nonmetric distinctions, decision trees work with data that can be measured in a variety of metric and nonmetric (quantitative and qualitative) ways. On the other hand, regression is a quantitative technique based on an approach to data manipulation that assumes that data values are linear and additive. Common intervals are inferred so that the metric distance between values 10 and 11 is the same as the metric distance between values 11 and 12, and so on. The statistical techniques at the core of the regression algorithm rely on data that is distributed according to the law of large numbers. When data deviates from this form, the regression technique begins to break down. Although there are techniques for handling deviations, the results become more uncertain and more difficult to produce.

Description of the Relationships As decision trees and regression have evolved, there has been an increasing cross-fertilization of techniques drawn from these two approaches. For example, regression approaches now accommodate nonlinear relationships and interactive effects in data. A common way to accommodate nonlinear relationships in a regression equation is to form a variable that is used to segment the data set so that different regression equations are fitted for different subsets of data. In this way, nonlinear effects are isolated by different regression equations that are formed for the different subsets. Decision trees can be used to construct the variable that is used to segment the data set. This is one of the many preprocessing functions that a decision tree can do before a regression analysis. SAS Enterprise Miner provides facilities to do this.

Likewise, decision trees have evolved to include simple and multiple linear relationships as splitting criteria in the construction of a decision tree. Decision trees can now apply fine-grained layers of branch partitions that are computed through resampling. As a result, the data space is cut into finer discriminations that resemble the discriminations made by regression equations. The differences between decision trees and regression have become smaller over the years.

Over time, hybrid approaches have evolved. A decision tree can be used as a preprocessor for regression to identify one or more atomic leaf nodes that, in turn, can become outcome groups to be modeled in the context of a regression analysis. By the same token, error-reducing functions, developed in the regression modeling framework, can be used to identify the attributes of the branches that should be used when creating the decision tree.

Because decision trees are developed recursively, they result in successively finer subcategories of data and successively smaller subgroups. Each subgroup is uniquely defined by the sequence of multidimensional branches that must be scanned to define the subgroup. These are the attributes of the subgroup that distinguish it and separate it from the other subgroups in the data. Unfortunately, as the decision tree grows deeper, it is harder to comprehend the overall view of data that the decision tree is describing. For this reason, decision trees that are more than three layers deep can be hard to understand. Also, as the decision tree deepens, branches that rely on fewer observations to determine their characteristics are identified and displayed. As a result, the reliability, accuracy, and reproducibility of the decision tree are threatened by the fewer observations that are used to shape the decision tree.

Regression techniques share a similar fate. Regression equations do not artificially divide the data into finer subcategories; however, they divide the data into finer functional descriptions. Ultimately, this means that regression equations suffer from opacity and a complexity of interpretation that decision trees do not suffer from. With decision trees, each and every subgroup can be precisely, easily, and uniquely defined through a visual or automated scanning of the decision tree or the rule representation.

Multidimensional Analysis with Trees

The ability of decision trees to include various combinations of inputs in the construction of a model of the relationships between inputs and targets makes it a first-class member of a broad family of powerful multiple term statistical models that includes multiple regression, factor analysis, and cluster analysis. In principal, decision trees can be extended to a multivariate form with multiple values in both the target and input positions. To date, no production versions of decision trees with multiple targets have been developed. But the operation of such a version can be produced in a limited but nevertheless extremely useful fashion. The construction relies on a technique that captures the multiple effects of multiple targets in one field. This one field can then be modeled using standard decision tree mechanisms as discussed here.

An Example with Multiple Targets

Hardware devices that are used for communications possess a wide variety of programmable hardware components, called slots. You can configure slots in multiple ways to deliver a variety of communications services such as file transfer, remote computer access, Internet page access, and voice-over-IP capabilities. The solution approach presented here is extracted from a presentation given at SAS Global Forum 2008 (De Ville, 2008).

It is extremely difficult to create high-level, summary descriptions of a collection of hardware devices. Any given hardware device could have one or more programmable hardware slots, and each slot can be any one of multiple family types.

Because of the many possible configurations, it is hard to identify common combinations that could be particularly useful in a given application area. Alternatively, it is difficult to identify common configurations that generate an inordinate number of faults, for example. This places a premium on deriving common configurations as an incentive to the dissemination of best practices on the one hand, and as a vehicle for superior quality control and trouble-shooting on the other.

A simple but effective summarization technique that accomplishes the objective of summarization is presented here. We show how over 6,000 different combinations of hardware configurations can be effectively summarized in only five clusters.

Problem Description

Table 4.2 illustrates the various family types that you can use in combination to characterize one particular type of communications device. These family types are fabricated, but an example label has been included as an aid to understanding.

Table 4.2: Distribution of Slot Type on Communications Hardware (Example)

HW Slot	Label	Frequency	Percent
SEC	Security	450	9.79
IP	Internet	26	0.57
VPN	Virtual Network	2235	48.62
WIRE	Hard Wire	83	1.81
MX	Multiplex	434	9.44
AIR	Wireless	19	0.41
AAIR	Advanced Wireless	1350	29.37

To grasp the magnitude of the problem, assume that a device can have as many as five slots and that each slot can be configured with any one of these seven hardware types. Many of these potential combinations exist in deployments of these devices internationally, but there are also common patterns that characterize the majority of the devices.

The problem is to discover these common patterns. Compressing or representing a multiplicity of possibilities into a smaller set of categories is a typical problem of dimensional reduction. Dimensional reduction problems are often solved through the application of numerical techniques such as cluster analysis, factor analysis, or principal components.

An alternative approach is to use Market Basket or Association Rules to identify the most frequently occurring combinations (or baskets) among the various 1-, 2-, 3-, 4-, or 5-slot devices. Association rules, however, do not identify common baskets according to the specific configuration of the hardware as is required here.

The alternative solution presented here relies on a technique that identifies common combinations in two ways:

- as one number that consists of a concatenation of all possible categories
- as a cluster formed through the application of a decision tree

Because many slot types are collapsed into one concatenated code, there is an immediate benefit of flattening multiple dimensions into one single dimension. This is equivalent to producing the full range of potential market baskets.

Secondly, a decision tree analysis is undertaken that uses the hardware configuration as the target or dependent variable. Clustering common combinations within various hardware configurations is then relatively simple.

Solution Approach

Create the hypercode. The solution begins with the creation of a single field, which is called a hypercode. The hypercode captures all of the unique, multiple hardware configurations that are possible, given specific combinations of slot types within hardware devices. This field is created by first sorting the characteristic slots that exist on any one hardware device in the collection, and then creating a multi-dimensional representation of the hardware slots that is based on a concatenation of all codes together.

Thus, if a particular hardware device contained three VPN slots, an MX slot, and an AAIR slot, then the hypercode would be as follows:

VPN+VPN+VPN+MX+AAIR

The table below provides an example in which we see that the Hardware ID is 120. There are five slots on this hardware device. Three slots are VPN types. Slots 4 and 5 are MX and AAIR, respectively.

Table 4.3: Example Source Record for a Hardware Device in the Analysis

Slot	1	2	3	4	5
Hw ID	120	120	120	120	120
Slot Type	VPN	VPN	VPN	MX	AAIR

To create a hypercode for this device, first sort the Slot Types. A discrete record is produced for each slot within the Hardware ID. Then a concatenated hypercode is built up across the individual records, and the final hypercode is written.

Table 4.4: Example Hardware Device Hypercode Derivation

	Hw ID	Slot Type	hyper	final
1	120	VPN	VPN	VPN+VPN+VPN+MX+AAIR
2	120	VPN	VPN+VPN	VPN+VPN+VPN+MX+AAIR
3	120	VPN	VPN+VPN+VPN	VPN+VPN+VPN+MX+AAIR
4	120	MX	VPN+VPN+VPN+MX	VPN+VPN+VPN+MX+AAIR
5	120	AAIR	VPN+VPN+VPN+MX+AAIR	VPN+VPN+VPN+MX+AAIR

The final record representation is shown in Table 4.5.

Table 4.5: Example Device with Hypercode

Slot	1	2	3	4	5	Hypercode
Hw ID	120	120	120	120	120	120
Slot Type	VPN	VPN	VPN	MX	AAIR	VPN+VPN+VPN+MX+AAIR

The code that produces this follows.

```
proc sort data=indata;
   by HWid SlotType;
run;

data indata_hyper;
   set indata;
length hyper $40;
   retain hyper;
   by HWid;
```

This creates a set of representative hypercodes that are based on the example slot families shown in Table 4.2.

This process of turning multiple hardware slots into a single vector representation takes a hardware entity with a multi-dimensional attribute (various hardware slots) and reduces the attribute to a single-dimensional form (the hypercode).

Cluster Similar Codes Together. The next step in the solution is to create a clustering of these hypercodes such that hardware devices with similar configurations are grouped together. Ideally, when such a cluster or grouping is formed, different configurations that are not a part of this initial cluster are separated into other groups, which, in turn, also represent devices with similar characteristics. The resulting clusters are as similar as possible, but they retain differences between the clusters.

The widely used technique of decision trees creates clusters based on a succession of variable inputs that are used as partitioning rules in the creation of branches and descendent nodes (or leaves). The typical approach begins with a root node and works through descendent nodes recursively until a tree-growth stopping rule is encountered.

A typical form for a decision tree is shown in Figure 4.13.

Figure 4.13: Top-Down Cluster Development in Trees

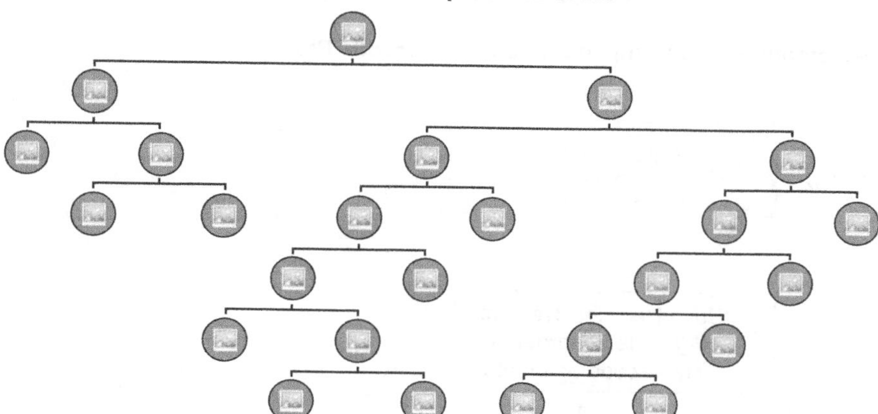

Here we see the root node at the top of the diagram followed by as many as six levels of binary partitions. Each of the terminal nodes – that is, the bottom nodes in the tree -- represents a cluster. You can find the attributes that define a cluster by tracing up the respective branches from the terminal node to identify the input fields that have been used as partitioning variables.

As shown in Figure 4.14, various forms of trees and therefore various forms of clusters can be produced, depending on whether binary splitting is used to identify the branches of the tree, or whether some k-way split is used. In the figure, splits are shown for 2-, 3-, 4-, 5-, and 6-way splits. The example below uses a 5-way split. This results in a very simple decision rule and correspondingly simple description of the hardware clusters.

Figure 4.14: 2-way and Multi-way Branches Illustrated

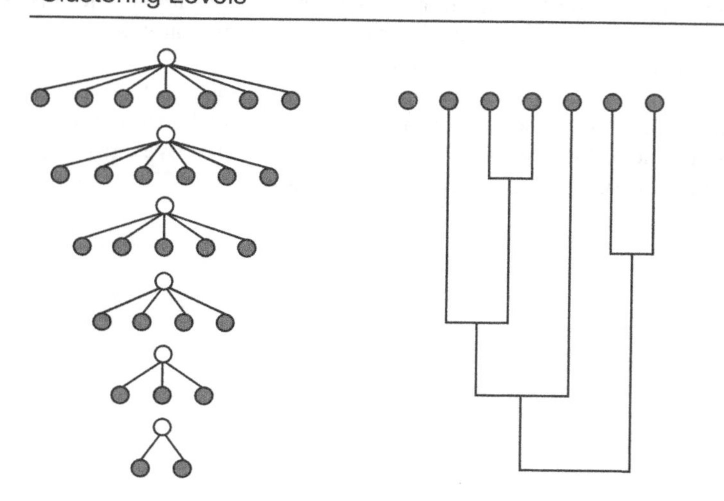

The final process combines concatenated hypercodes and decision trees, as shown in Figure 4.15.

Figure 4.15: A Schematic of the Complete Cluster-Hypercode Process

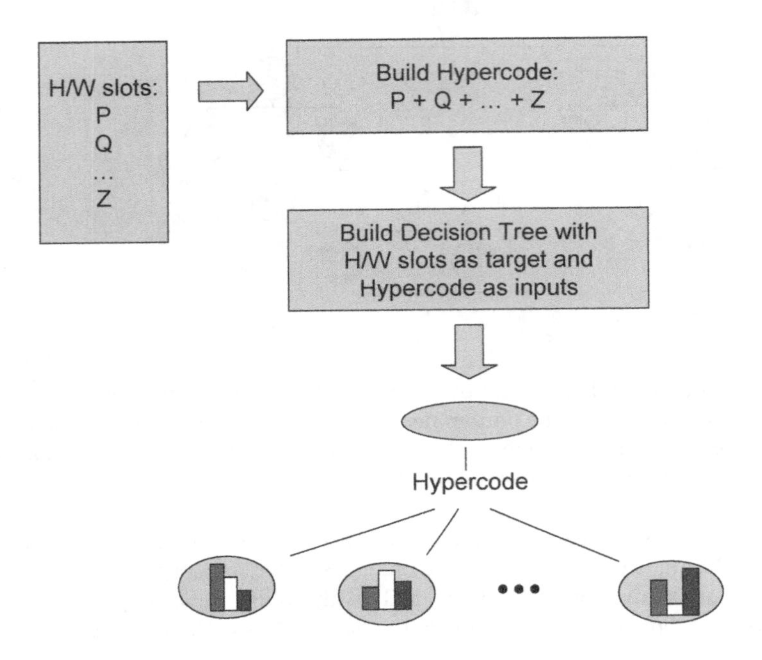

In the example analysis presented here, the unclustered hardware slots, which are shown in Table 4.2, are used as the target value in the analysis. The associated hypercodes are used as inputs. Because we know what the individual slots for a given site are, we can determine what combinations of slots (hypercodes) tend to be associated with a given configuration of slots.

The overall results are displayed in Figure 4.16. Because we use a 5-way branch partition on the first input, we need to partition the root node of the decision tree only once: we derive five clusters (or five terminal nodes) with one single partition on the hypercode input variable.

Figure 4.16: Top Level of the Decision Tree Showing Five Baseline Clusters

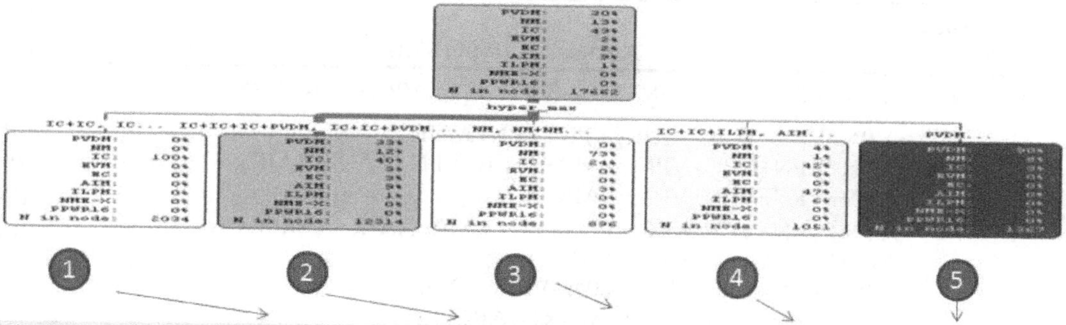

	1	2	3	4	5
PVDM	0%	22%	0%	4%	20%
NM:	0%	12%	73%	1%	8%
IC:	100%	40%	24%	42%	3%
RVM:	0%	3%	0%	0%	0%
KC:	0%	3%	0%	0%	0%
AIM:	0%	9%	3%	47%	0%
ILPM:	0%	1%	0%	6%	0%
NMK-X:	0%	0%	0%	0%	0%
PPWR16:	0%	0%	0%	0%	0%
N in node:	2034	12314	896	1051	1367

The attributes of the five descendent nodes are shown in the following table. Terminal node contents or clusters are described moving from left to right.

Table 4.6: Detailed Contents of the Clusters (Terminal Nodes)

Node (Cluster)	Associated HyperSlots	Modal Categories
1	VPN+VPN, VPN, VPN+VPN+VPN, VPN+VPN+VPN+VPN	VPN (100%)
2	VPN+VPN+VPN+AAIR, VPN+VPN+AAIR, VPN+VPN+VPN+VPN+AAIR+AAIR, VPN+AAIR+AAIR, VPN+VPN+VPN+VPN+AAIR, VPN+VPN+VPN+MX, VPN+VPN+AAIR+AAIR+AAIR+AAIR, MX+MX+AAIR+AAIR+AAIR, VPN+VPN+VPN+VPN+VPN+VPN+MX+MX, VPN+VPN+VPN+AAIR+AAIR+AAIR, VPN+AAIR+AAIR+AAIR, EVM+VPN+VPN+MX+AAIR+AAIR, MX+MX+AAIR, SIP+AAIR+AAIR, VPN+VPN+VPN+VPN+MX, VPN+VPN+VPN+VPN+VPN+VPN+VPN+VPN+VPN+MX+MX, VPN+MX+AAIR+AAIR+AAIR+AAIR+AAIR, VPN+VPN+VPN+VPN+VPN+VPN+VPN+VPN+VPN+VPN+MX, SIP+MX+MX+MX, MX+MX+AAIR+AAIR, VPN+VPN+AAIR, VPN+VPN+VPN+MX+AAIR+AAIR+AAIR+AAIR, VPN+VPN+VPN+VPN+MX+AAIR+AAIR, VPN+VPN+AAIR+AAIR+AAIR, IP+IP+EVM+AAIR+AAIR, SIP+SIP+AAIR, SIP+VPN+MX+MX, IP+IP+IP+IP+EVM+EVM+AAIR, EVM+VPN+AAIR, IP+IP+VPN+VPN+MX+AAIR, SIP+SIP+EVM+AAIR	AAIR (33%) VPN (12%) MX (12%) SEC (9%)
3	MX, MX+MX, VPN+MX, VPN+VPN+MX, MX+MX+MX, SIP+MX+MX, SIP+MX+MX+AAIR, VPN+MX+MX+MX, VPN+MX+MX, MX+MX+MX+MX, VPN+VPN+VPN+VPN+MX+MX+MX	MX (73%) VPN (24%)

(continued)

Node (Cluster)	Associated HyperSlots	Modal Categories
4	VPN+VPN+WIRE, SIP, SIP+VPN+VPN, SIP+VPN+VPN+VPN, SIP+SIP+VPN+MX+AAIR, VPN+VPN+VPN+WIRE+AAIR, VPN+WIRE, WIRE, SIP+VPN+VPN+VPN+VPN, SIP+SIP+AAIR+AAIR, SIP+VPN+VPN+VPN+VPN+AAIR+AAIR, VPN+VPN+VPN+WIRE, SIP+SIP+VPN, SIP+SIP+MX+AIR+AAIR	SIP (47%) MX (42%)
5	AAIR, MX+MX+AAIR+AAIR+AAIR+AAIR, AAIR+AAIR, VPN+MX+AAIR+AAIR, MX+AAIR+AAIR+AAIR+AAIR+AAIR+AAIR, VPN+MX+MX+AAIR+AAIR+AAIR+AAIR+ AIR, AAIR+AAIR+AAIR, MX+AAIR+AAIR+AAIR+AAIR, MX+AAIR+AAIR+AAIR, MX+MX+MX+MX+AAIR+AAIR+AAIR+AAIR, MX+MX+AAIR+AAIR+AAIR+AAIR+AAIR, AAIR+AAIR+AAIR+AAIR, MX+MX+MX+AAIR+AAIR+AAIR+AAIR+AIR	AAIR (90%) MX (9%)

Cluster Descriptions

Cluster 1 presents a simple result: when hardware devices consist of a varying number of VPN slots, then the slot combinations themselves are a series of VPN slots. In the results, we can see that 1-, 2-, 3-, and 4-way concatenations of JD slot types are presented.

Cluster 2 is more interesting. This set of communications devices contains mostly expanded wireless AAIR (33%), VPN (12%), MX (12%), and SEC (9%) slot types (the percentages show the relative percentage of each of these slot types across the entire family of slot types available in this cluster). This means that a third of the slots in this set of hardware devices are AAIR slots. The frequencies for the other slot types indicate that VPN, MX, and SEC slots are present (approximately 10% each). One sequence of hypercodes (VPN+VPN+VPN+MX+AAIR+AAIR+AAIR+AAIR) indicates that as many as eight slot locations are present among these devices.

If we assume that we could configure each of these slot types two or three ways, in various sequences, across a total of eight possible slots, it is easy to see that over 10,000 combinations are possible. The decision tree results shown here indicate that these types of communications devices tend to have as many as 24 configurations, as shown in Table 4.5. The most frequent combination, shown first in the list, is VPN+VPN+VPN+AAIR. So we could characterize Cluster 2 as a predominantly VPN service device, potentially for expanded wireless applications.

Clusters 3, 4, and 5 resolve to 10, 14, and 13 main configurations, respectively. As with cluster 1 and 2 results, we observe a significant reduction in the total possible number of configurations.

Tractable Results

The results shown here indicate that extremely complex hardware configurations can be effectively summarized using a hypercode concatenation technique coupled with decision trees. In this case, we are able to present thousands of combinations in five clusters. The clusters contain 1, 24, 10, 14, and 13 combinations, respectively; 62 combinations in total.

In Cluster 2, you saw that with eight possible slots and four primary hardware components (AAIR, VPN, MX and SEC), there were millions of potential combinations, but only 24 combinations were identified in the cluster. The most common combination, VPN+VPN+VPN+AAIR, could be used for targeted marketing, to suggest best practices for this type of hardware device.

Chapter 5: Theoretical Issues in the Decision Tree Growing Process

Introduction

The discussion in previous chapters has served to highlight and illustrate the two major characteristics and attractions of decision trees as methods of dealing with data:

- to extract and apply information from data, particularly predictive information
- to extract and communicate insight from data

The first major characteristic, prediction, emphasizes the accuracy and reproducibility of the decision tree model and does not emphasize the underlying form, structure, or intrinsic comprehensibility of the decision tree. The second major characteristic emphasizes pattern detection, identification, and communication. This approach is exploratory and can be used as a precursor to other techniques, such as multidimensional cube reporting or building predictive components.

The strength of decision trees for exposition is in the decision tree's ability to uncover multiple effects, both visually and intuitively. As is often the case in the application of statistical results to problem solving, there can be a trade-off between the form of a tree that is suggested by business-rules versus tree results that are based on the strict application of numerical results. One big advantage of decision trees is that they enable you to quickly model various scenarios that blend numerical results with business rules and experience. The construction of an optimal tree result that blends both business rules and numerical efficacy is more of an art than a science. We'll propose some guidelines in the following discussion that might help you develop your own approach.

Since the initial substantial deployment of decision trees as data analytical tools in the '70s and '80s, decision trees have solved many of the early problems that labeled them as ineffective prediction tools in terms of validity, accuracy, and efficacy. Recently, encouraging developments have suggested that predictive goals and descriptive insight goals are not necessarily antagonistic. Results presented by Breiman in the area of random forests (Breiman, 2001) and Friedman in gradient boosting (Friedman, 1999b) demonstrate the effective communication of what seems to be deeply complex, potentially obscure multi-tree models. Both random forests and gradient boosting are multi-tree approaches that resample the analysis data set numerous times to generate results that form a weighted average of the resampled data set. When summarized, the results of many decision trees are better than the results of a single decision tree. In both random forests and gradient boosting, the emphasis is directed away from detecting the form and representation of a single decision tree, toward presenting a graphical representation of the associated decision tree predictive components such as scores and deviations.

Whether the goal is prediction or insight, a major benefit of decision trees is exposing relationships and patterns in data, generating predictive results, and communicating these findings. The user's task is to understand the various approaches and to choose wisely when conducting analyses.

Crafting the Decision Tree Structure for Insight and Exposition

Here we talk about the art of growing a decision tree for insight (extracting conceptually appealing information from data) and exposition (displaying the decision tree results in a form that communicates insight and informs policy and planning). The goals of insight and exposition differ and complement the goal of using decision trees to extract key relationships and predictive structure from data (which satisfies the requirement of maintaining an overall form, structure, and sequence of branch formation in the decision tree).

You might find it useful to think in terms of telling a story when growing a decision tree to reveal information and communicate results. The storyline and theme needs to support the conceptual framework of the audience. The story illuminates key interests and potentially contains a few plot twists that upset conventional ways of looking at the data and therefore pave the way for the development of insight and improved understanding.

In telling the story, it is important to have a beginning, middle, and end. The story should be told in terms that are familiar to the audience. And, while it can be useful to include a few twists in the plot, the insights that are revealed should be plausible. The best way to ensure a good story line is to construct the decision tree in-line with the conceptual model of the area that the decision tree is designed to illuminate. For example, if you are looking at purchase behavior, then the attributes of the decision tree need to reflect concepts that are relevant to purchasing behavior. If the application is quality control and you are looking at part failures, then the attributes of the decision tree need to reflect concepts that are relevant to part failures.

Every application area in which expository decision trees can be deployed is characterized by concepts that either explicitly or implicitly exist in the minds of the audience. Concepts have been measured and reflected by different entities in the data set and can be linked differently, particularly if the entities suggest different links based on the empirical characteristics of the data. However, there is always an underlying story line, a presumed relation, and a presumed cause and effect or sequence of causes and effects. Some decision trees are more comprehensive than others. One characteristic of a comprehensive decision tree is that the data in the conceptual area that is being explored contains a range of related attributes. As a result, the story that is told by the decision tree reflects both a plausible set of relationships and a fairly complete set of relationships (i.e., to the extent possible, the substantial drivers of the relationships being explored have been included).

To build this type of decision tree for exposition, the following tasks should be performed:

1. Define the business and/or scientific question.
2. Determine the main features of a conceptual model that describes the major constructs involved in the question resolution.
3. Determine the data measures, fields, and field values that will become the operational components of the conceptual model when the model is translated to form the decision tree.
4. Develop the story line (i.e., the presumed sequence of events as the operational components unfold to tell the story).
5. Determine key relationships or potential plot twists to be examined in shaping the form of the decision tree.
6. Translate the tree results into a form that illuminates the original question.

Conceptual Model

So far, the decision tree concept has been described by the analogy of telling a story. It can also be described by comparing it to the traditional scientific method. In the scientific method, you begin with a theoretical model. From the theoretical model, you build an exposition that consists of an operational model. Then, you form constructs that reflect empirical data and that can develop a set of hypothesized relationships and proposed tests of hypothesis that can demonstrate the mirroring between the operational model and the theoretical model.

While the decision tree is constructed to be readily consumed and useful to the audience, it is also constructed to support scientific rigor and defensibility on scientific and engineering grounds. This kind of defensibility is essential because ultimately, the decision tree results are designed to become enterprise policy deployments that produce substantial savings and profits. Therefore, no matter how appealing the story is or how compelling the plot is, the results must be scientifically robust in order to support a review and implementation program that could offer significant benefits.

A decision tree can illustrate the operation of the model used to explain the relationship being examined. This is especially true when you want to expose the relationships between the multiple factors in the model and the target. In explanatory modeling, the most important consideration is crafting the structure and sequence of the branch partitions used to expose the model and explain the results.

Consider the following weather forecasting example. This example will show how the local, observed weather conditions result from a stream of prior events that can be measured and used to explain variations in local weather.

Figure 5.1: Illustration of a Conceptual Model of the Analysis of Weather Patterns

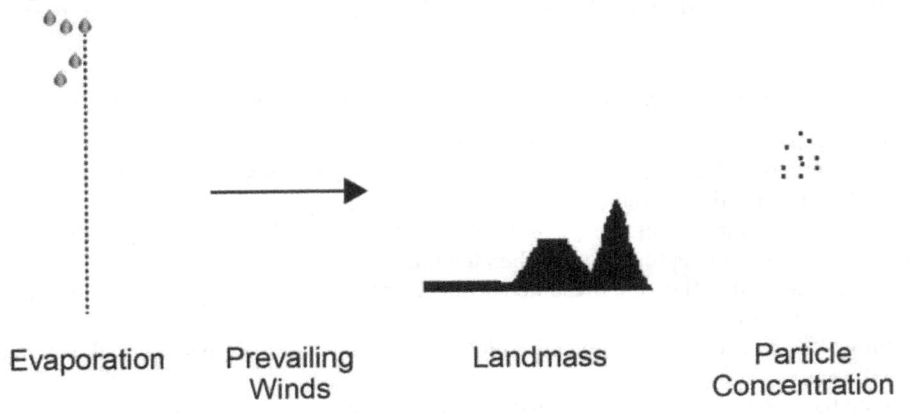

A simplified model of weather might be as follows—water evaporates over the ocean, forming clouds of moisture in the air. These clouds move in a direction based on the prevailing winds. Moisture can be squeezed out of the atmosphere depending on what happens to the prevailing winds as they encounter landmasses. Finally, if there are particles of dust in the air, the condensation of water droplets to produce rain is accelerated.

The prediction of rain or shine depends on the cumulative operation of a sequence of variables. The winds have no effect unless there is evaporation for them to carry. Landmasses and particle concentrations on their own cannot produce rain; they need the winds that carry moisture. And, without landmasses and particle concentrations to run into, moisture could be carried by the winds forever and never be released.

If there is rain and a mountain range nearby (a frequent event on the West Coast of North America), you should not conclude that the mountainous surroundings have produced the rain. This is an example of a spurious relationship. To know the whole story, you need to measure the evaporation over the ocean, the speed and direction of the prevailing winds, and so on.

When growing the decision tree so that it reflects this order in the model of the domain, you should observe this sequence of events. There are many competing branches at the top level of a decision tree. It is common to have a number of alternative branches—all significantly related to the target (as indicated by their logworth measure). The question becomes, "Which branch should I select?" For the purpose of exposition, it is best to select branches in the order that conforms to your modeling framework. This usually involves selecting a branch with a lower logworth than other available branches. In the weather example, this suggests that the topmost branches of the decision tree should reflect evaporation rate. Next, you should select branches reflecting the operation of the wind. Then, you should select branches reflecting landmass profiles and particle concentrations, in that order. In some cases, you might select a branch that does not pass a test of significance. It is important to include these nonsignificant branches in the displayed decision tree so that the conceptual model is properly reflected.

These tasks can be called the analytical approach or the analytical framework. They are illustrated in Figure 5.2.

Figure 5.2: Illustration of the Top-Down Work Breakdown to Develop an Analytical Model

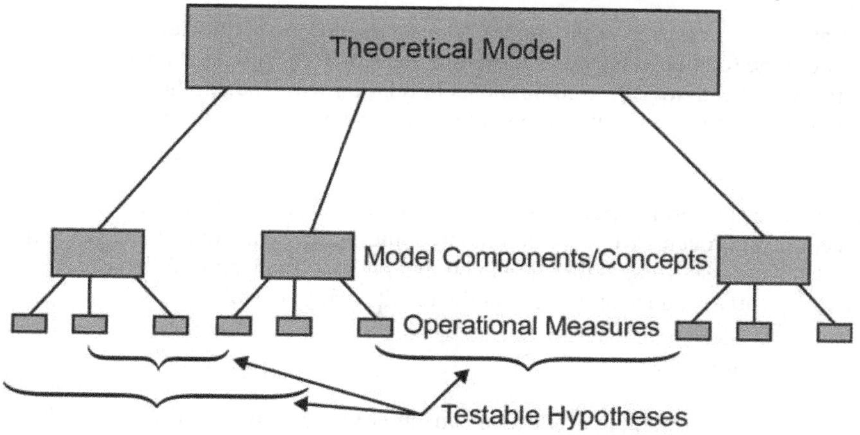

In the purchasing example developed in Chapter 2, the theoretical model could presume that purchase preference follows socioeconomic status and is influenced by life cycle factors such as marital status, home ownership, and children. Other influential factors can be presumed, such as sociopsychological lifestyle and personal preference. A predisposing factor such as a recent loan or home purchase might be a signal that indicates an immediate or impending change in one or more of the dimensions of life cycle. To better understand these concepts, consider the balloon-and-line diagram in Figure 5.3.

Figure 5.3: Illustration of an Analytical Model

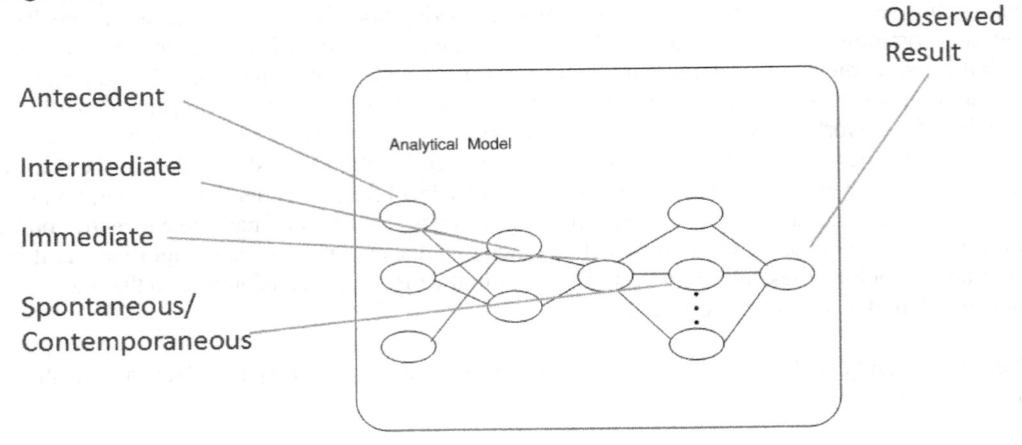

The balloons illustrate the components of the model. The connecting lines illustrate the presumed relationships. To model the relationships in the form of a decision tree, you need data that encapsulates the presumed behavior of the components in relation with one another. The data is used to derive the empirical measures that will operationalize the behavior that you want to examine.

As the model evolves, you can form hypotheses to test as you construct the decision tree that represents the conceptual approach. After forming the hypotheses, you can control and guide the construction of the decision tree. This, in turn, puts you in a position to meaningfully, constructively, and efficiently produce and interpret the results that you find.

Following is an example hypothesis:

Is there a limit to the variety of drinkable products a consumer will be attracted to? How is this limit distributed in my current (and future) market population? How will this vary over time and by customer?

Question 1: How can various products be clearly packaged and differentiated?

Question 2: Which of these product packages can give me a satisfactory ROI given the competitive environment and satisfactory marketing mix (product, price, place, promotion)?

Each layer in the model can consist of theoretical constructs that are operationalized by the database contents. Operationalizing is the process of assigning numeric tokens for conceptual entities. Table 5.1 shows an example.

Table 5.1: Theoretical Constructs

Socioeconomic Factors	Socio-psychological Factors	Psychological Style	Range of Preferences	Buying Decisions
Age	Recreational Activities	Outward-Directed versus Inward-Directed	Outdoor Products	Buy Product X
Education				
Marital/Family Status	Type of Employment		Indoor Products	Buy Product Y
Income	Indebtedness			
Antecedent	*Intermediate*	***Immediate***	*Spontaneous*	*Result*

The hypothesis is that the propensity to buy the explosive, multi-fruit blend (Product X) is a function of adolescent, upwardly mobile, outdoor-oriented, outward-looking types with elevated levels of spending power.

To support the evolution of the plot as the story unfolds while building the decision tree, you must introduce the branch partitions in the order that is implied in the conceptual model shown in Figure 5.3. The top branches of the decision tree are grown to reflect the background socioeconomic factors. Next, socio-psychological factors and psychological style factors are introduced. Finally, the range of preferences is introduced (if available). Preferences provide the most accurate and attention-grabbing presentation of the combination of factors and relationships that are optimal in capturing the data in the context of extracting meaning to understand purchase behavior.

This method of growing the decision tree can correctly preserve sequence information and explicit and implicit time-ordered relationships in the data so that the results reflect the logic of preconditions and consequent targets. Simpson's paradox states that if you are examining relationships among the independent variables, then the sequence of the construction of these variables can be important. There are limits to computation-based measures; this is an opportunity for the analyst to rely on knowledge of theory, practice, and experience when choosing the branches and sequence of branches in the decision tree.

Furthermore, in a data mining context, it is common for a data set to have hundreds of potential inputs serving as explanatory or descriptive decision tree components. It is not sufficient to let the decision tree grow in a way in which branches are picked on the basis of predictive strength. To explain the results, it is important to construct the form and sequence of the branches. A heuristic approach that is based on the underlying conceptual model can substantially reduce the construction burden on the analyst.

A practical method is to use a fishbone or Ishikawa diagram. In the retailer data first discussed in Chapter 1, you can start with an Ishikawa diagram as shown in Figure 5.4.

Figure 5.4: Illustration of an Ishikawa Diagram to Organize the Constructs of an Analytical Model

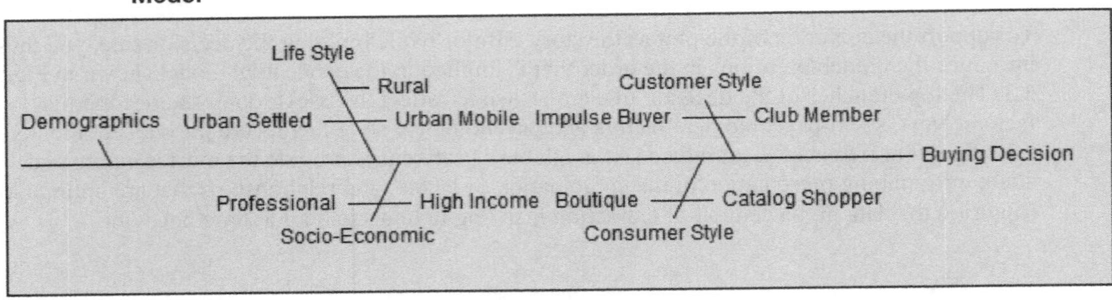

In practice, determine what dimensions are likely to be relevant by examining the source data. The source data is shown in the following figure.

Figure 5.5: Example Variable List and Associated Operational Dimension

Number	Variable	Dimension
1	gender	Demographic
2	occupation	Socioeconomic
3	owns_truck	Lifestyle
4	owns_motorcycle	Lifestyle
5	owns_RV	Lifestyle
6	valueOfCar	Socioeconomic
7	length_of_residence	Socioeconomic
8	maritalStatus	Demographic
9	age	Demographic
10	hasBankCard	Consumer Style
11	hasStoreCard	Consumer Style
12	has_card	Consumer Style
13	has_credit_card	Consumer Style
14	has_upscale_store_card	Socioeconomic
15	children_home	Lifestyle
16	adultsInHH	Lifestyle
17	income	Socioeconomic
18	has_new_car	Socioeconomic
19	Recency	Customer Style
20	lifeTransactions	Customer Style
21	lifeVisits	Customer Style
22	NetSalesLife	Customer Style
23	state	Demographic
24	bathroomPurchases	Customer Style
25	bedroom	Customer Style
26	kitchen	Customer Style
27	juvenile	Customer Style
28	table	Customer Style
29	windowDisplay	Customer Style
30	couponPurchase	Customer Style
31	Monetary Value	Customer Style
32	Frequency	Customer Style

There are a number of potential inputs for the five dimensions, as shown in Table 5.2.

Table 5.2: Measurement Inputs

Dimension	Measures (Inputs)
Demographic	4
Socioeconomic	6
Lifestyle	5
Consumer Style	4
Customer Style	13

There are numerous potential combinations of branches if all measures for all dimensions are inserted into the decision tree. By constraining the order of entry, the potential number of combinations is reduced considerably. This might be considered a shortcoming if the only goal in growing the decision tree is predictive accuracy. However, it is damaging beyond repair if the goal is to grow a decision tree that can be displayed and explained in conceptual terms that are relevant for an audience that is eager to better understand its business or research.

In summary, remember that data represents operational measures of concepts or of analytical constructs. For example, atmospheric pressure is a concept. The height of a column of mercury is used as a method of making the concept operational. Here, as we climb up the side of a mountain, the atmospheric pressure diminishes. If we carry a barometer with a column of mercury in it, we will see the column of mercury diminish as we move up the mountain. When we have a good operational construct, it tends to replace the original concept. So, for example, in meteorology the term “millimeters of mercury” or “millibar” is now synonymous with atmospheric pressure.

In all models, data is used as an operational measure of some analytical construct. The characteristics of the phenomenon being modeled can be captured and exposed by examining the relationships between the data points that operationalize the various terms in the construct. To maintain the time-ordered nature of effects and produce a more readily interpretable decision tree, you should introduce branches with fields that move from the left to the right in the causal sequence of effects (as you move from the top to the bottom of the decision tree).

Predictive Issues: Accuracy, Reliability, Reproducibility, and Performance

Decision trees are sensitive to the sequence of branch growth. Once a branch is selected, it affects the structure of the entire decision tree below it. Thus, it is critical to be very careful in selecting the sequence of branches that are introduced into the decision tree if the goal is to interpret the decision tree components to gain a better understanding of the factors that influence the area under examination.

If the goal is raw prediction, then the sequences of chains of branches in the decision tree in terms of their expository value are less important. What is important is identifying sequences that have predictive value. In this situation, decision tree quality relates to how well the decision tree performs in terms of accuracy, reliability, and reproducibility. Because some computations can be time-consuming, quality also relates to the ability of the decision tree to deliver results within defined time periods. A number of techniques are available to capture these various dimensions of quality.

Sample Design, Data Efficacy, and Operational Measure Construction

Sample Design. The issue of sample design applies to all situations where empirical data is used to gain knowledge of the environment. The data that is used to gain this must be representative of the environment. For example, with lemonade stand sales, you assume that the data is representative of the situation that you are modeling. There are random elements in data collection. For example, different people pass the lemonade stand on different days; thus, the data collected depends on what day the data was collected.

The variability and potential gaps in data can affect the results produced in the decision tree. Suppose lemonade sales are modeled on either time of day (morning versus afternoon) or on location (corner versus mid-block). On some days, such as hot days, time of day might best explain sales. On other days, location would provide a better prediction. If data were collected over several days, including both hot and not-so-hot days, the time model and the location model might predict sales with about the same degree of accuracy.

For prediction, either model will produce good results. However, for explanatory purposes, the models might suggest completely different things. One model might suggest selling only in the afternoons, saving the cost of morning operations. The other model might suggest paying more for a corner location. It might turn out that an underlying variable—in this case, temperature—would explain both the time-of-day and the location differences, but because of the variability in data collection, this variable is missing from the data.

Data Efficacy and Operational Measures. In addition to constructing a comprehensive data sample, it is important that the data contains information that relates to factors that are known to be or are likely to be relevant to understanding and predicting the target. So, if temperature is relevant, then you need to have measurements of temperature.

It is also important to have true metrics to reflect measures. Cold, warm, and hot might not be as good as temperature measured by a thermometer when building the effects of temperature into the model. Many measurements of human behavior do not follow a linear form. As amply illustrated by extensive work in psychometrics, most forms of human perception are not only nonlinear, but are often non-monotonic and circular in nature.

Strong Sets of Predictors. After sample design, data efficacy, and measuring, the most likely issue to emerge in growing the decision tree for prediction involves identifying strong sets of predictors. A predictor is strong if it is consistently and accurately related to the target or outcome under examination. In a previous example, height and weight were generally good combined predictors of gender. In a decision tree framework, you would identify ranges of height as nodes in a decision tree that have an associated likelihood of predicting male or female gender. The same is true of weight and height-weight combinations.

The difficulty with decision trees is selecting ranges that work well. A range of 90–110 pounds in one sample might be almost exclusively female. Yet, in another sample, this range might be 50% females and 50% males. The challenge is to find not only strong predictors, but also strong ranges or cutpoints in the branches of the decision trees. This is the rule, rather than the exception. This is especially true in regression when some of the input variables are associated or collinear. Variations in the cutpoints of some variables can suppress the effects that might otherwise be observed at lower parts of the decision tree. This suppression effect can lead to unstable models with substantial variations in accuracy and explanation when we attempt to generalize the results.

Another challenge is to find strong combinations of predictors (i.e., height-weight combinations that produce consistent predictions of gender over time). Tree-based models are particularly unstable. Small changes in the training data can completely change the structure of the decision tree. If the variable in the first splitting rule changes and another branch is substituted, then the descendent branches can be very different.

The two most generally applied approaches to identifying predictor combinations in decision trees rely on statistical tests of significance, usually with multiple Bonferroni adjustments and various forms of validation or cross-validation.

There are many options for dealing with instability in the inputs that are selected:

- use stand-in variables (a variable that approximates the unstable variable, but is more reliably related to the outcome)
- create composites (e.g., principal components or factor scores, or another reduction measure that is more stable because it represents the weighted combination of multiple indicators)
- get more data to capture more power in the relationships that are being examined

A best-fit model is neither too big nor too small. At some point in the growth of the decision tree, after one or two top predictors have been identified, you might be tempted to introduce another predictor. For example, in the height and weight data, you might want to use age to capture another dimension of the predictive space and, consequently, produce a more accurate or reliable prediction.

How large should the final decision tree be? The interpretability of the decision tree decreases as you descend to its lower levels. And, the statistical power of the results is weakened. The developers of the CRT approach have done the most work in this area. The CRT approach grows an overly large decision tree and then prunes.

As shown in Figure 5.6, there is a point of divergence between the readings provided by a training data set and the readings provided by a test data set. At the point where the training form of the decision tree model is not replicated in an independent sample, it is time to prune the decision tree.

Figure 5.6: Illustration of Drop in Accuracy Using a Resubstitution Statistic

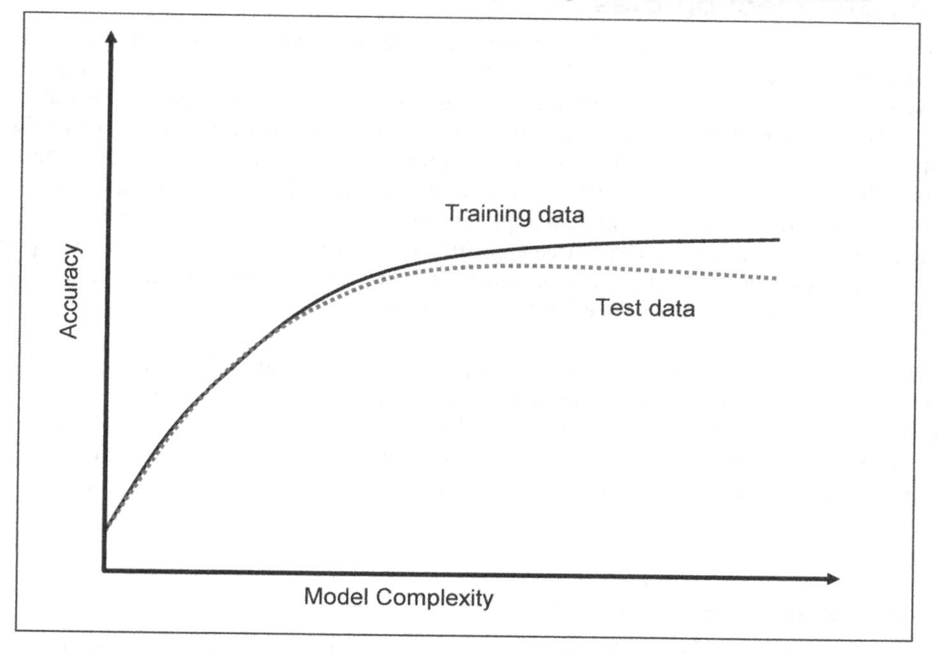

With decision trees, the accuracy calculation depends on the level of measurement of the target. With categorical targets, accuracy is the total number of correct class predictions that are given by the tree over the total number of classes in the data set. For continuous targets, the accuracy rate is measured by the deviation scores of predicted versus actual values.

Choosing the Right Number of Branches

If the goal is predictive, then there are rare cases in which an unordered, multi-way split can create better predictive results than a binary decision tree. If you have strong theoretical reasons for presenting an unordered grouping of codes in predetermined (or computed) clusters, then you should. You might even get superior predictive results.

If the goal is explanatory, there is no statistical answer, per se. You can use experience or theoretical reasons to assist you in creating multi-way groupings that are logical or that concisely capture the nature of the subpopulation. This will help you interpret and communicate the results. It will enable you to find more meaningful interactions among the predictors. If you have reasons for a binary decision tree (e.g., North-South, East-West), then use one. If you have reasons for a multi-way decision tree (e.g., East-Mid-West-West), then use it.

Perspectives on Selection Bias

The early versions of decision tree approaches to statistical decision making were both praised and cursed almost from the beginning of their early adoption beginning in the late 1960s and early 1970s. One seminal article, by Morgan and Sonquist (1963), demonstrated the then-uncanny ability to detect localized, variable interaction effects in regression models. Following close on the heels of these early demonstrations were criticisms of the variable selection properties of the decision tree products at the time. One notable article by Doyle and Fenwick (1975) demonstrated how variable selection bias undermines the apparent applicability of the decision tree method. The authors demonstrated how iterative decision tree algorithms could bias the selection order of inputs so that weak, even meaningless, inputs were selected as branch splits when better and stronger inputs were either overlooked or were placed in the lower ranks of precedence.

Major efforts were underway during this formative time to address variable selection bias. One corrective approach was proposed and published by Gordon Kass (1975). This approach proposed chi-squared adjustments to the usual AID approach and was consequently named CHAID. The CHAID approach proposed adjustments to the level of significance of the tree branch splits. These tests, while approximate (and sometimes overly conservative), directly addressed the issue of variable selection bias.

Description of Variable Selection Bias

In principle, the goal of unbiased variable selection is simple: at each step in the tree-growing process, the algorithm should select the best, unbiased input to form the next branch of the tree. This selection can be reliably made when, for example, an input (called Z) is predictive of a target Y and an input (called X) is not. In this simple case, the algorithm should select Z as the input to form the next branch.

We quickly find that there are difficulties even in this rather simple situation. For example, in any given collection of data, we recognize that we will likely be operating on a sample which we probably assume is randomly drawn. So we recognize a special case of a specific sample where the variable X happens to be more strongly associated with Y because of chance variation. The algorithm would then legitimately—but erroneously—select X as the input split. What we require here is a long-run perspective in the algorithm, so that our software would recognize that, over successive samples of data, Z is more strongly associated with Y than with X and would then select Z as a better partitioning input. In this view, we expect our algorithm to be able to select the better input more often.

As we look at sources of selection bias, we consistently find that some inputs are more susceptible to chance variation than others. Generally, the more choices for splitting, the more risk that chance variation in a sample will defeat the strategy that is in place to choose the best variable for a split. Nominal inputs with many categories, often referred to as high-cardinality categorical inputs, allow the search routines greater freedom to segregate the observations into branches. This greater freedom leads to a greater number of potential splits. More potential splits increases the likelihood that an unusual or atypical split will show up in the algorithm with a strong effect by chance alone. These kinds of fields are therefore susceptible to variable selection bias.

Potential Remedies to Variable Selection Bias

Let's explore three strategies for overcoming biased input selections:

1. **Selection based on comparisons to a common (null) distribution.** This discussion reviews the approach that employs a comparison of the live data with various random data distributions. This process identifies the most likely input as the input that is least associated with the random data.
2. **Selection based on p-values of a permutation distribution.** Instead of using a null distribution from a random population, this approach assumes that the data is fixed as given, and creates a null permutation distribution by permuting the values of Y.
3. **Selections based on pre-computing a measure of association.** Here we pre-select the splitting variable based on a simple association measure. The examples illustrate that this approach works quite well.

The Null Distribution

How does the null distribution depend on the number of categories, the number of observations, and the proportions of the classes of a target value Y? A common strategy for fairly comparing different inputs is to first compare each individual input with random data. In this strategy, the worth of the best split using an input X is compared to the distribution of worth values obtained from repeatedly generating and splitting random data. The variables in the random data are the same type and have the same number of nominal categories (as in the example of X input and Y target). The proportion of worth values from the random data that is larger than the split worth using X is called the *p-value* of the *null hypothesis* that the input and target are unrelated. The distribution of random worth values is the *null distribution*. The input with the smallest p-value is selected for splitting. This strategy is designed to minimize the selection bias. In the following examples, the null distribution acts as a surrogate for potential bias. The best input, chosen from a number of inputs, is considered to be the input with the lowest relationship with this surrogate, or null, distribution.

As an example of how this procedure might work, assume that we have a binary response, or target, called Y. This binary target has equally likely values and has an unrelated nominal input variable X(b) (with b values also equally likely). At this point, we can see that the data has no signal and is pure noise. Even so, let's investigate the distribution of the worth of the best split. To do this, we simulate an example by drawing a sample of N=500 observations. We find the best split (using the Gini criterion), and save the split worth that was identified.

We do this 1,000 times, and display the distributions for situations where the b values in X were equal to 2, 14, and 30 categories. The results are shown in Figure 5.7. Notice that a thorough search almost always finds a split of some ostensible worth, even in pure noise. The more categories, the more splits are available, and the greater the quantiles in the distribution of worth. These results directly correspond to our earlier observations concerning the origin of bias.

Figure 5.7: Null Distribution of Gini Reduction (N=500, 1,000 iterations)

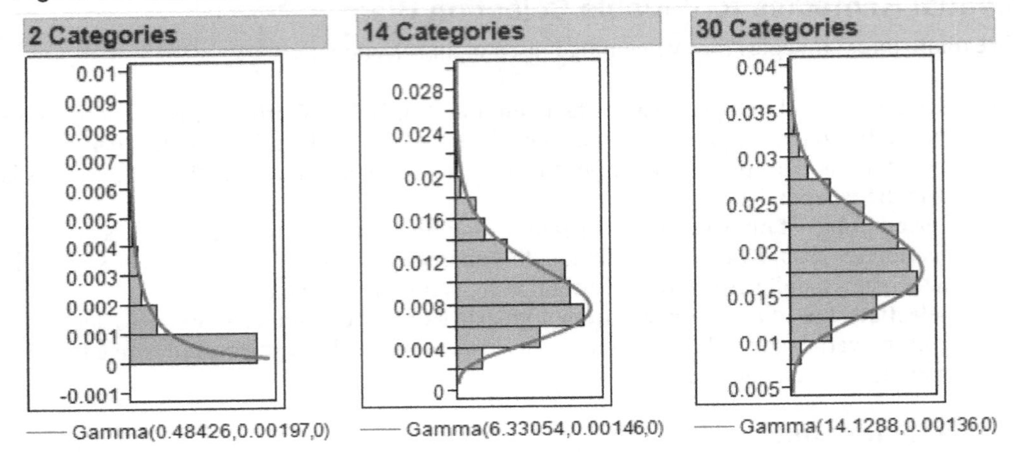

A gamma distribution fits these results well. The gamma distribution is the most common distribution used to approximate skewed continuous data—data that looks lopsided with the bulk on one side and a longer tapering tail on the other. The normal distribution is symmetric and therefore not skewed. The NULL distribution of split worth values is skewed. Overall, we find that a gamma distribution fits the null distribution quite generally, regardless of the number of categories, the number of observations, the ratio of target classes, or the splitting criterion (in addition to Gini, we could use Entropy, or the p-value of a chi-square distribution). Other researchers have noted this in different contexts (Dobra and Geryke, 2001, and Sall, 2002). It appears that we could use a gamma distribution to calculate an unbiased test statistic. As shown in Figure 5.7, what is required is to determine the two parameters of the gamma distribution. These parameters depend only on the mean and variance of the worth of a split.

Variations of the Null Distribution

The null distribution varies with the number of input categories and the number of observations. Figure 5.8 shows box plots for 1,000 data sets for each of ten values of the b categories of input X (2, 6, 10, ..., 38). The median worth of the best split trends up linearly with b. For b=6, the median is about 0.0035. For b=38, it is 0.0248. The average worth, though not shown in the plot, also trends up linearly.

Figure 5.8: Distribution of Split Worth by Number of Categories of X = 2, 6, ..., 38. N=500

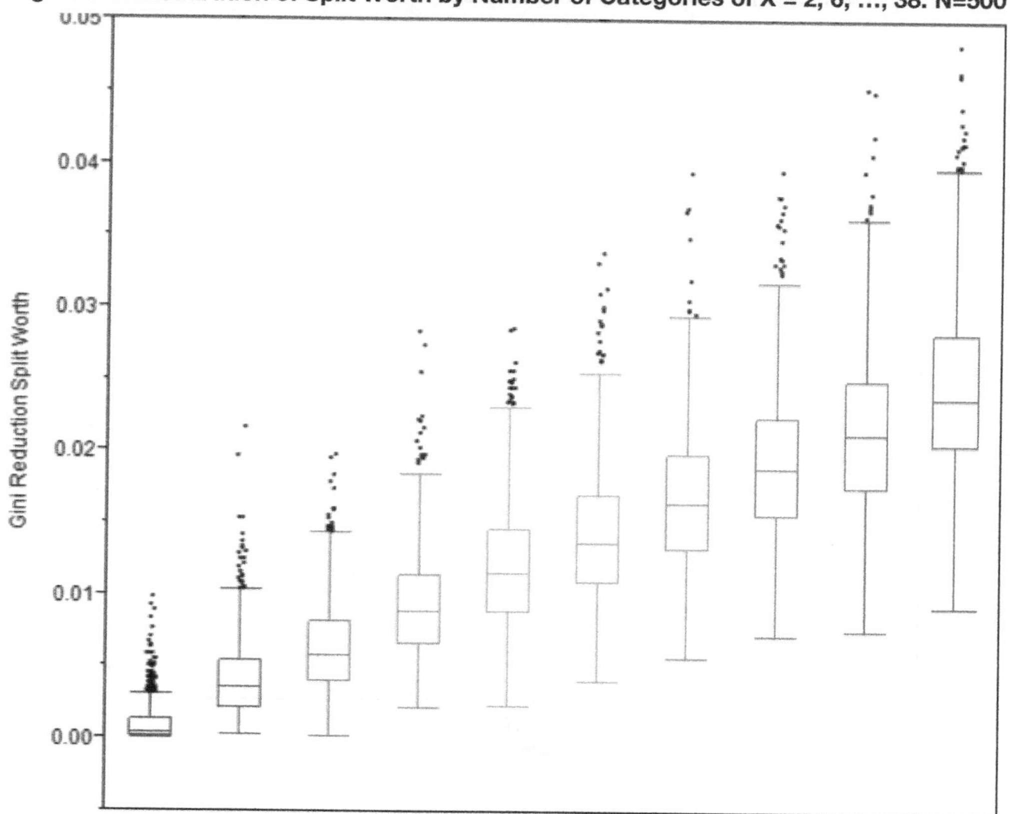

Figure 5.9 shows the effect of varying the sample size. This figure shows the median worth for each of ten values of b and four values of N (500, 1000, 2000, 4000), with four lines resulting. The top line represents N=500 and depicts the medians of the box plots from Figure 5.8. Increasing the number of observations N causes the trend to flatten out. The slopes of the lines are 0.000642, 0.0003215, 0.00164, and 0.0000812. Notice that the slopes are proportional to 1/N. For a fixed b and N, the median worth is therefore proportional to b/N in these examples. These results were replicated for the mean value (results not shown here).

Figure 5.9: Split worth versus Number of Categories for Various Sample Sizes

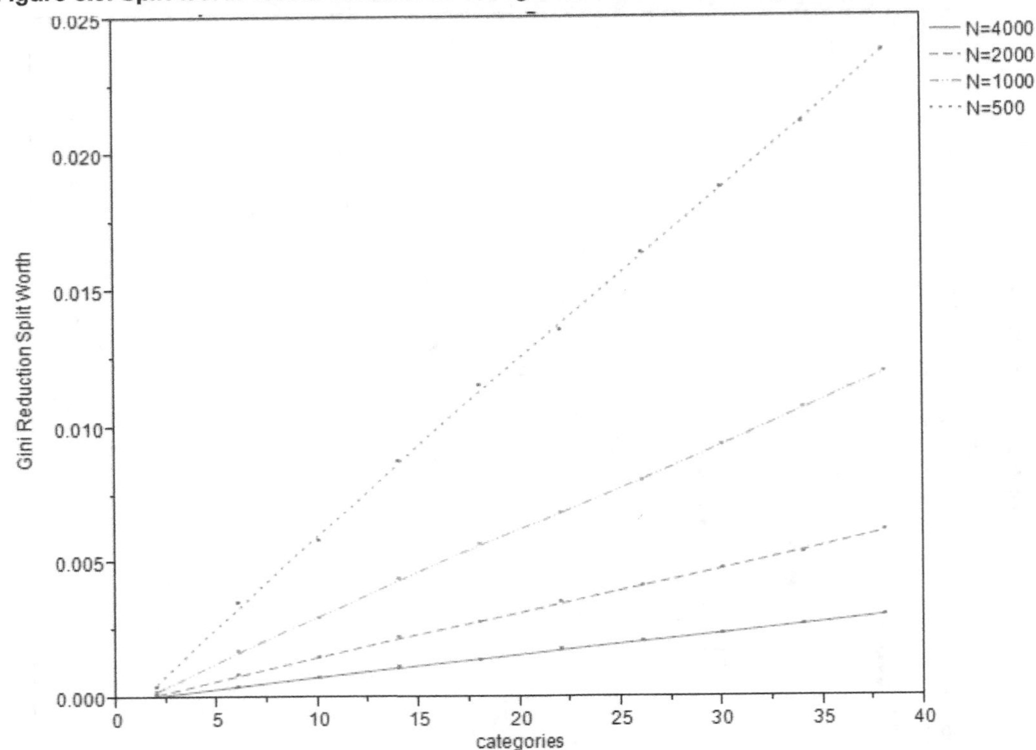

Variations of the Null Distribution with Different Proportions

So far we have shown situations where the target is binary with equal class sizes. Here are results where the proportions in the target classes vary, the number of categories in the input varies, and the sample size varies. In the examples, p denotes the probability that Y equals 1 (assuming Y takes values 0 and 1). In the examples shown so far in this section the target was binary; therefore, p equals 1/2.

For purposes of this example, a value V is computed as p(1-p)b/N.

The top graph in Figure 5.10 plots the median worth from the Gini criterion versus V for seven values of p, ten values of b, and four values of N. The values of p are: 1/2, 7/16, 3/8, 5/16, 1/4, 1/8, and 1/16. There are 7 by 10 by 4 = 280 points, and they fall very nearly on a line. The regression R-square equals 0.99992. The linear fit is still strong, regardless of whether we use another quantile instead of the median (such as the upper quartile). A plot of the upper quantile would show that the points spread out slightly, mostly at small values of V.

The close association of median worth and V results only when the Gini criterion is used to compute worth. The other three graphs in Figure 5.9 plot median worth against V for Entropy, PCHISQ, and CHAID. PCHISQ denotes the p-value of a chi-square statistic. It is the same as CHAID, except the worth using CHAID equals the worth using PCHISQ divided by the Bonferroni factor, the number of possible splits. The results suggest that only Gini has a clear relationship with a target of varying proportions (illustrated by varying the number of categories in the input and the size of the sample used for computation).

Figure 5.10: Median Worth versus (b p(1-p)/N) for Different Split Criteria

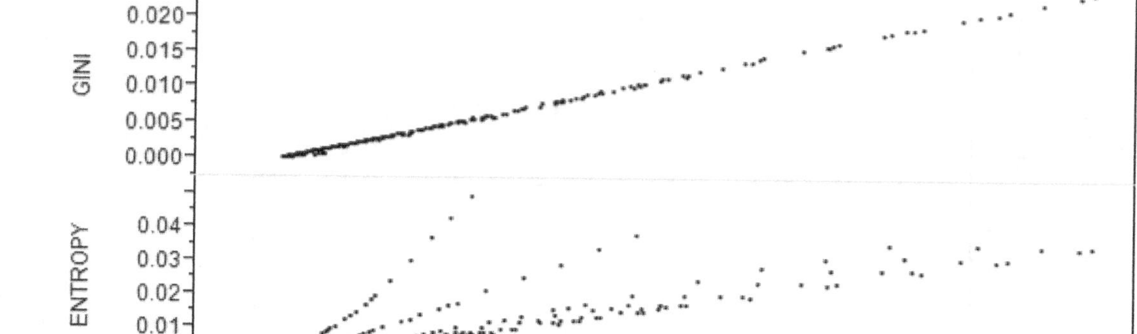

Figure 5.11 is the same as Figure 5.10 except the horizontal axis is now b, the number of categories of X, instead of V. Notice that the median PCHISQ worth has a consistent linear relationship with the number of categories. Notice also that for PCHISQ and CHAID, the median worth of the null distribution does not get smaller as the number of observations gets large.

Figure 5.11: Median Worth for Various N and p Values

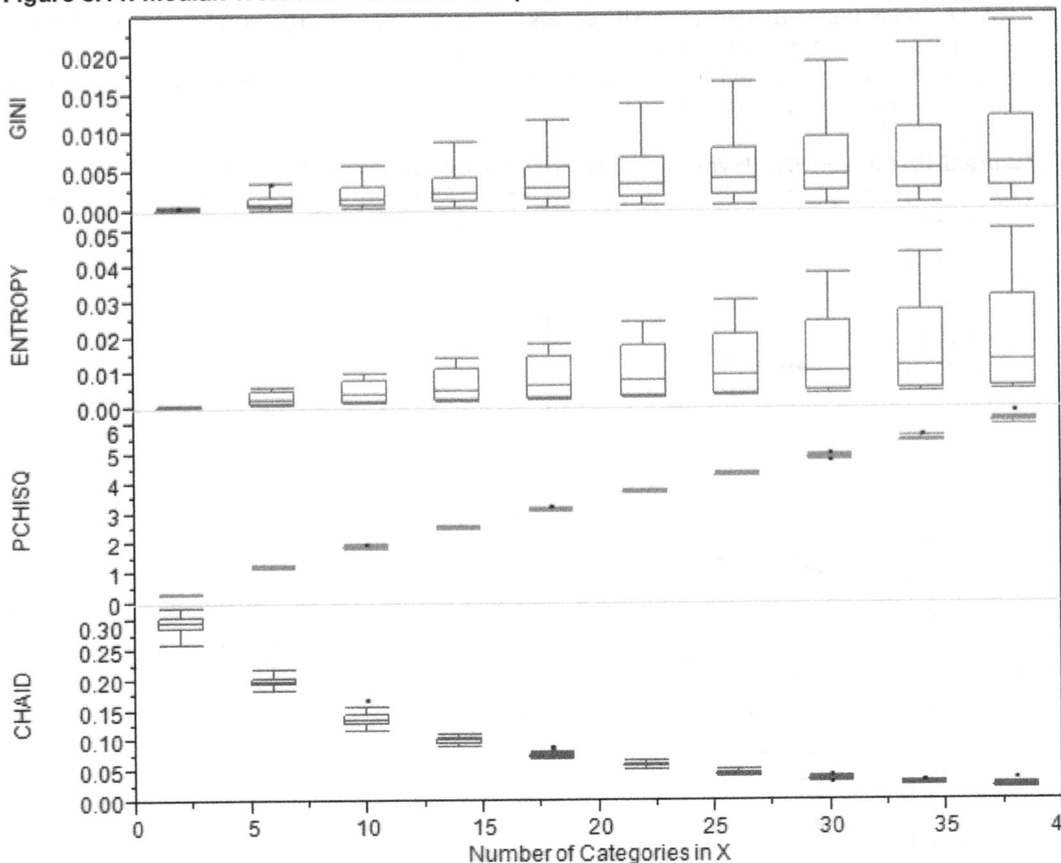

In summary, when an input variable is unrelated to the target, a split search almost always finds some split, even though it is spurious. The distribution of worth values of these spurious splits appears to fit a gamma distribution. Knowing the mean and variance of the worth values of the random splits, the appropriate gamma distribution is known, and selection bias is minimized using the p-values from these gamma distributions.

The figures show a relationship between the median worth and other factors that is very different for different splitting criteria. However, given a criterion, the plots are suggestive that the median, and hence the mean, are predictable. For Gini reduction, the relationship between the average or median and b, N, and p looks especially simple. Other authors have computed preliminary, but so far only partial, results. As these results mature, we expect to be able to use these relationships to compute a bias adjustment for input selection.

More than one researcher has expressed the mean and variance of the null distribution as a function of the number of categories and other factors. Then, taking the null distribution as a gamma distribution, as shown here, they apply the common strategy of using p-values to compare inputs, and have a variable selection method that works well in the situations that have been examined.

Alin Dobra and Johannes Gehrke (2001) derive the mean and variance for Gini reduction for a nominal input and target, and are satisfied that their procedure minimizes the bias. However, they assume a multiway split into b branches. Their expression for the mean does not seem to apply to binary splits and therefore does not apply to all situations discussed here.

John Sall (2002) produces an example solution that models the mean and variance by fitting a neural network to training examples. He uses a splitting criterion proportional to N times Entropy reduction. The model appears to work well in some situations and is used in JMP.

The Null Permutation Distribution

Gordon Kass (1975) was the first author in the field of decision trees to try to derive an expression for the null distribution as previously described. His approach differed from the approach above: Instead of generating random data, Kass uses the given observations in the data set and permutes the target values. For each permutation, he finds the worth of the best split. The null permutation distribution is the distribution of these worth values over all permutations.

Instead of actually permuting and splitting the data, he presents formulas for the permutation distributions and solves them numerically for a few simple cases. These simple examples do not generalize well. So, for example, the formulas developed by Kass are not accurately solvable for these examples, even 35 years later.

Kass moved on to develop a practical implementation of his theoretical notions, resulting in the development of CHAID. Kass was certainly aware that using Bonferroni adjustments in CHAID would result in a conservative test of significance as a split criterion (Kass, 1975). This means that some potential splits are rejected because of an overly conservative test. The test becomes more and more conservative as the number of categories increase, particularly for categorical inputs with many combinations of categories to consider. At the time of the development of Kass's approach, it was normal to use single-digit categorical fields (primarily due to limited computer resources). For this reason, Kass did not explore the extreme conservatism that crept into his test adjustments as the number of categories in the fields in the analysis increased (particularly in non-metric, "free" fields [as Kass described them]).

The permutation approach pioneered by Kass still has advocates. Anne-Laure Boulesteix has used it and has published some results, for example (Boulesteix, 2006). The reported results do not claim to handle more than five categories. Matching procedures have been created and are packaged in R. A method for a nominal input and binary target has been implemented in the exactmaxsel package (Boulesteix, 2009). The authors know of no software, including exactmaxsel, that can find the exact permutation p-value for the examples shown here (our examples include 2, 6, 10, up to 38 categories).

On the other hand, we re-ran some of our examples using permuted data instead of generating independent samples as discussed previously. A gamma distribution fit the distributions of split worth values from the permuted data sets as well as they fit the independent samples. The means and variances were also close. These results might indicate that a permutation approach is warranted only on smaller data sets.

A Study of Pairwise Bias

You can understand the influence of bias by studying what happens when two variables compete with one another in a given predictive situation. In this illustration, assume that Z is a predictor of a target, called Y. There is also another weakly associated predictor of Y, called X. If X is a categorical variable that can contain various numbers of categories, then by chance we will observe relations between some ranges of categories in X and the target Y. In various scenarios with different numbers of categories for X, it is likely that more relationships will be observed as the number of categories increase.

To illustrate, let Z(c) denote a nominal variable with c categories and with some predictive relation with Y, and let X(b) denote a nominal variable with b categories independent of Y. For fixed b and c, the pairwise bias is defined as the proportion of samples in which a splitting procedure selects X(b) instead of Z(c). This is bias because we select X, even though it is not related to Y, because of a chance artifact that is a result of the number of categorical values that are contained in X.

In a particular sample, chance variation might result in X(b) being more strongly associated with Y than Z(c), or in a subtly different fashion, X(b) might facilitate a split with a greater worth than Z(c). Ideally, an unbiased selection process would pick Z(c) in most random samples and would not succumb to the chance variations that occur in X(b). Variable selection is based on the worth of the split unless stated otherwise.

Study Data. Assume that Y has values 0 and 1 with equal probabilities. In line with the approach taken by Kononenko (1995), we define a joint distribution of Y and Z(c). For Z(c), assign half of the c categories to a set C(0) and the other half to set C(1). For a category in C(0), the probability that Y equals 1 is one half. For a category in C(1), the probability is 0.6. In the ideal situation of infinite data, the best split on Z(c) will separate C(0) from C(1), regardless of c. The worth of the best split on X(b) goes to zero as N becomes large, and therefore the bias will go to zero as N becomes large.

Figure 5.12 is a matrix of scatter plots comparing the Gini worth of Z(c) versus X(b) for b and c equal to 6, 22, and 38 categories. Each plot contains 1,000 points, one for each simulated data set. Each data set contains 4,000 observations. Within each plot, points below the diagonal represents data sets in which X(b) is selected instead of Z(c). The proportion of such points is an estimate of pairwise bias. For example, the lower left plot has no points in the lower diagonal, signifying that Z(6) was selected over X(6) in all 1,000 data sets. The matrix of plots shows that the bias increases from left to right and top to bottom. This is expected because the chance of a spurious split increases with the number of categories for fixed N and p.

Figure 5.12: Gini Worth of Z(c) versus X(b), X(c) (6, 22, and 38 Categories)

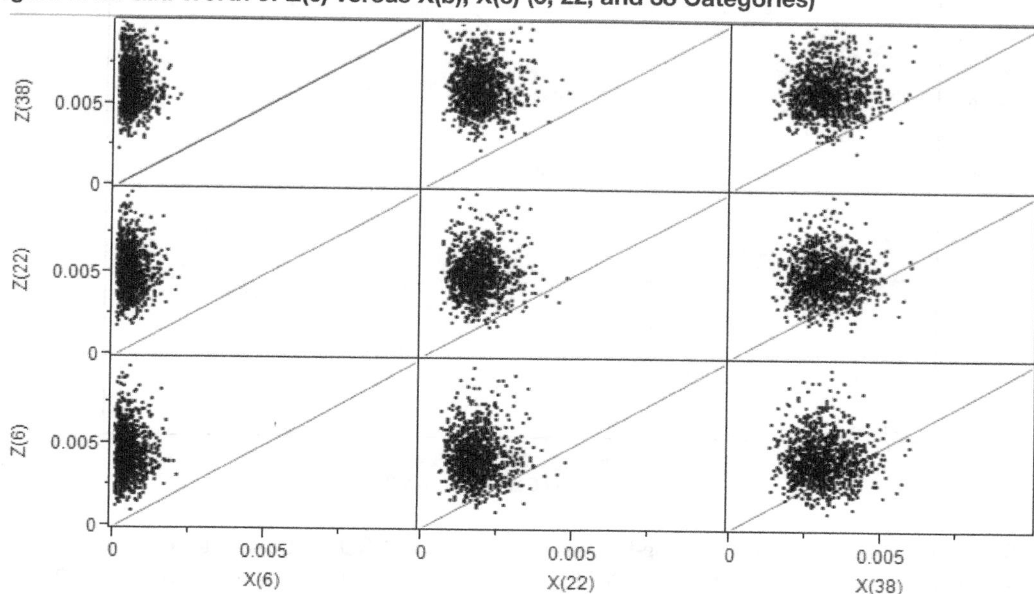

Selection bias also decreases as N decreases. Figure 5.13 shows the proportion of samples in which X(b) is selected over Z(c) for N = 500. As in Figure 5.12, the bias increases with b and decreases with c. However, the bias is much larger now. The proportion of times X(6) is selected instead of Z(6) is about 0.24, instead of 0. When comparing X(b) and Z(c) with the same number of categories, the bias increases with the number of categories. With 38 categories, the bias is 0.38. The signal gets lost as the number of categories increase and the number of observations remains the same.

Figure 5.13: Proportion of Times Gini Selects X(b) Over Z(c). N=500

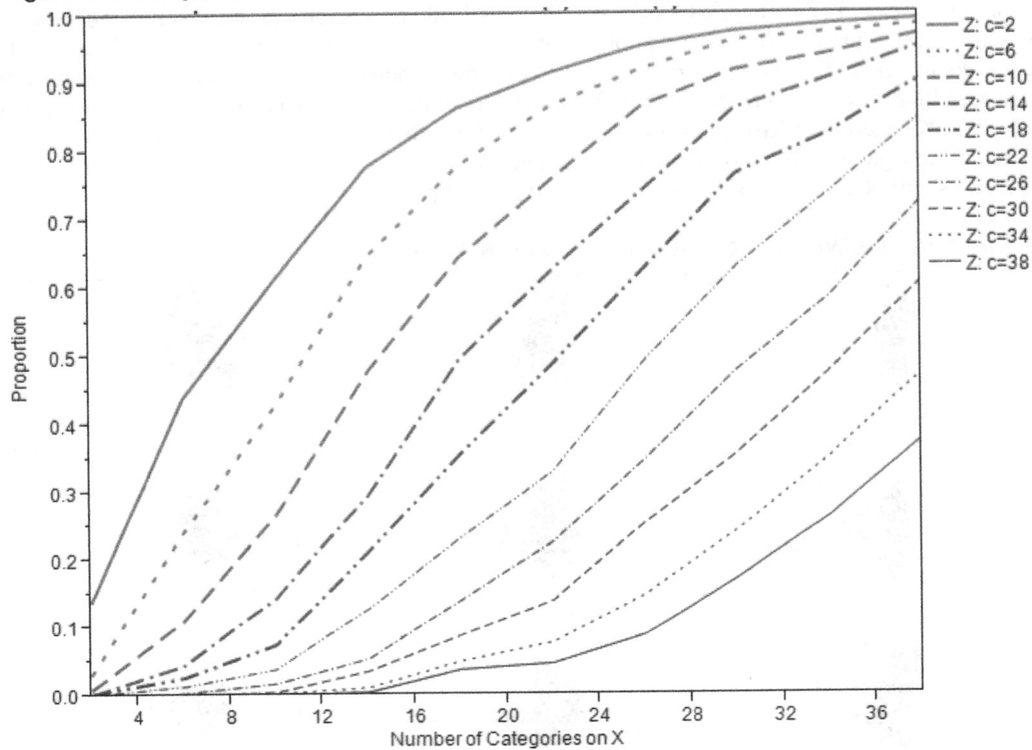

The same results occur when using the Entropy or the PCHISQ splitting criterion. This result is somewhat surprising because Figure 5.11 shows the splitting criteria to have very different relationships between median worth and the number of categories.

Figure 5.14 shows the results using the CHAID criterion (which uses a Bonferroni adjustment for the number of categories). The results are reversed from the previous plot in two ways: The bias now decreases with b and increases with c. The line representing Z(38) is now on top and Z(2) is on the bottom. CHAID adjusts too much for the number of categories. As previously mentioned in the discussion of the null permutation distribution, Kass was well aware this would happen.

When comparing X(b) and Z(c) with the same number of categories, the bias from CHAID is the same as from the other criteria. As a proof, consider that W(x) and W(z) denote the worth values using the CHAID criterion with X(b) and Z(b) respectively. It follows that the worth values using PCHISQ are B times W(x) and W(z), where B is the Bonferroni factor for b categories. X(b) is selected with CHAID when W(x) > W(z), which happens exactly when B W(x) > B W(z)(i.e., when X(b) is selected with PCHISQ).

Figure 5.14: Proportion of Times CHAID Selects X(b) Over Z(c). N=500

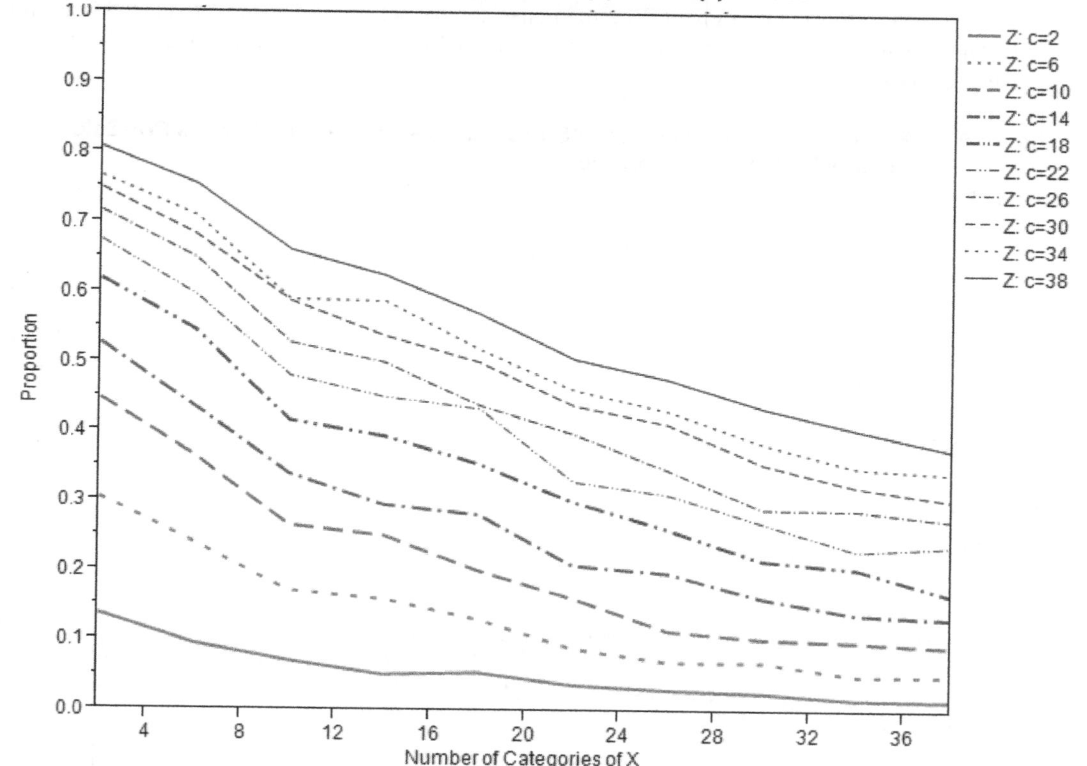

Pre-Selecting the Splitting Variable

The QUEST algorithm was a follow-up development to CRT that was co-developed by Wei-Yi Loh (1997), a former student of Leo Breiman. Wei-Yi Loh had been sensitized to the outstanding issue of variable selection bias, and so he set out with co-developer Shih to address it. The algorithm that resulted produces negligible bias.

The Loh and Shih idea is to first choose an input variable using a statistical test, and then search for a split on that variable and only that variable. QUEST does not use the worth of a split to select the splitting variable. However, it does use a Bonferroni adjustment, which is conservative. The test also has two stages, and involves a Bonferroni threshold to switch between stages. It looks complicated compared to what came nine years later from Hothorn, Hornik, and Zeilus (2006). These authors changed the two-stage test in QUEST to an asymptotic permutation test of association. This approach works well. In addition, this algorithm easily adapts to new data types such as censored variables.

Figure 5.15 shows the results for 500 observations using an implementation in a new SAS procedure, HPFOREST. For a given Z(c), the pairwise bias between Z(c) and X(b) is the same for all b. The bias increases with c, and converges around 0.35. For 2,000 observations, the bias converges around 0.13.

Figure 5.15: Results from Experimental SAS Procedure—Proportion of Times Pre-Selection Selects X(b) Over Z(c). N=500

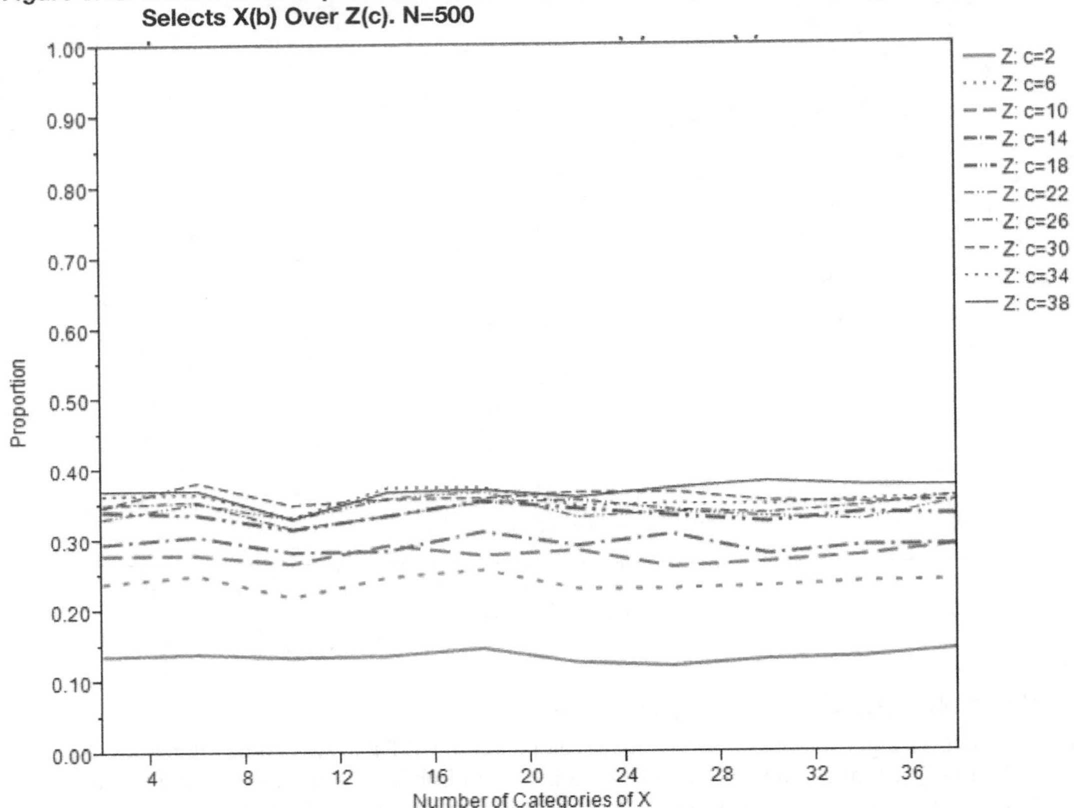

Conclusions

Variable selection bias will occur from chance variation in the data. Variables that provide many split choices, notably nominal inputs with many categories, are the most likely to allow the most opportunity for an algorithm to find a misleading split. When Gini and Entropy reduction are used as split criteria, the bias decreases with the number of observations. CHAID overly penalizes variables with many categories. Only a restrictive set of examples was considered here: binary target, nominal input uniformly distributed, and binary splits. The intention was to demonstrate the issues.

Efforts to reduce the bias have gone in three directions:

- finding the mean and variance for a null gamma distribution
- finding an algorithm for the null permutation distribution
- selecting a variable before searching for a split

Each of the three methods has achieved some success. At the moment, the third method—selecting a variable before doing the split search—appears to be the most generally applicable approach. Hopefully more will come from the other directions as future developments unfold.

Multiple Decision Trees

A central concept of statistics is that, in general, an average of many observations is a more reliable estimate of a future observation than any given single observation is likely to be. And so it is with decision trees. In medical diagnoses, it is common to ask for a second opinion. Even though the primary diagnosis comes from an expert in the field, this expert might have idiosyncrasies that can bias the outcome of the medical examination. Similarly, a single decision tree might reflect idiosyncrasies that can bias the predicted results. A prediction from numerous trees is like a majority opinion from a committee of doctors: the idiosyncrasies of a minority of doctors are either over-ruled or discarded.

The numerous trees that are grown and averaged to produce a composite score can be selected and grown autonomously as a parallel process or sequentially. Autonomous trees use different bootstrap samples for the training data. The trees are different because their training data are different, and only because their training data are different: the trees are created identically in all other respects.

Figure 5.16: Illustration of Sequential Boosting and Autonomous Bagging Ensemble Processes

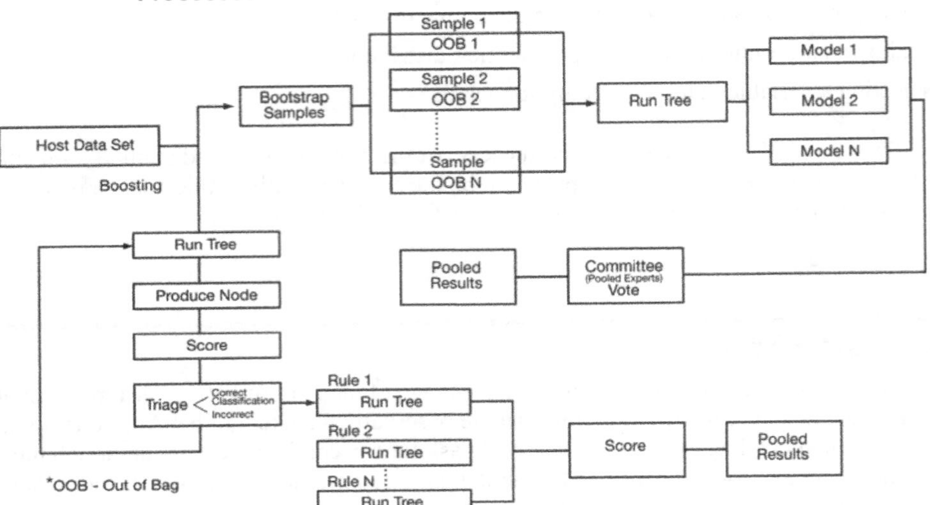

The common methods of producing ensembles are illustrated in Figure 5.16. The approach of taking various autonomous samples is illustrated in the top part of the diagram. A sequential approach is shown in the bottom part of the diagram. In the sequential approach, multiple trees are grown in sequence so that different approaches to the data can be applied in successive steps. Downstream steps can take advantage of earlier results; for example, observations that are not well predicted in earlier trees are manipulated to increase the estimation ability of the new tree in the sequence. In the final stage, the results from the various sequentially developed trees are averaged to produce the shared estimate. Intuitively, you might sense that the ability to manipulate dependencies from step to step can produce highly effective predictive models that promise to produce the best overall estimates of any approach.

Ensembles

The aggregation of multiple trees to produce a composite result is called an ensemble model. This leads to the definition of an *ensemble* as a model with a composite prediction that combines the estimates from a number of simpler models, sometimes called *base learner*. Ensembles of decision trees are now widely accepted as being better for prediction than a single decision tree.

The ensemble forms its prediction either by averaging the predictions of the individual component models or by taking a vote based on the execution of the component models against a data set. For continuous targets, the prediction is always the average. For classification (categorical) targets, the predicted class is the one most voted for, i.e. the one that most of the individual models predict. The predicted probability of a class is the proportion of models that voted for that class. Alternatively, the probability of a class can be defined first as the average probability of the class among the individual models. The ensemble then predicts the class based on these averaged posterior probabilities. The alternative ways of defining predictions for an ensemble produce different results. Different methods might be more successful in certain situations. In the long run, it is difficult to recommend any given method because all methods have their virtues and appear to work equally well in a wide variety of circumstances.

Ensembles are created by designing a strategy to operate on different trees that are drawn from the same training data set. The method to be used in producing a composite, average score is also defined. One simple way to construct various instances of different trees is to train the trees with different samples of the available data. The first tree ensembles that did this imitated the statistical practice of bootstrapping by sampling with replacement. Like bootstrapping, this enabled estimates of the distribution of a statistic.

As ensemble methods have evolved, so too has the notion of building separate training instances by using sampling without replacement. Accordingly, many ensemble methods now prefer sampling without replacement.

Ensembles smooth out the predictions. Smoothing works especially well for a technique such as decision trees that dices data in a recursive, piecewise linear fashion (i.e., the decision boundaries in the data are formed by linear cutpoints that are determined by the values on the left and right sides of the branch partitions). Thus, a decision tree with 20 leaves partitions the data space with 20 linear edges. These 20 linear edges partition the data space into 20 rectangles. The 20 rectangles are sub-segmented by lower partitions in the decision trees that are grown in each of the descendants of the original 20 leaves. Bagging 100 twenty-leaf decision trees will average 100 step functions across the data space. This process forms a more continuous set of boundaries in the data space and approximates the data-fitting characteristics of techniques such as regression and neural networks.

In general, bagging smooths and blurs the normal, hard-edged partitions that are formed by a single decision tree. If the true relationship between the inputs and the target is not well represented by a single, hard-edged partition, then smoothing may help reveal the relationship. One side effect of smoothing is that it obscures the overall structure of the decision tree. Because the bagging approach takes the average of many decision trees, the original, readily interpretable decision tree structure of any one of the single components of the bagged result is easily lost.

An ensemble of autonomous models is successful if the base learners have two properties: they must be good and they must be different. The task of finding a good base learner is the traditional one of finding a good model for the data. The task of finding a mechanism to generate variants of a model, all of which are good, is the central task of building an ensemble of autonomous base learners. In order to generate depth and range in the predictions, consider only mechanisms that inject random changes in the model building, either in the training data or in the algorithm.

A modeling algorithm that produces a very different model in response to a random change is said to have large variance. To produce a good model, an algorithm with large variance must resemble the training data well. Such a model is said to have low bias. The variance of an ensemble of autonomous models is smaller than the variance of any of its base learners. The bias of the ensemble is the same as the bias of a base learner. Thus, the reason an ensemble of autonomous models is better than an individual base learner is because the ensemble reduces the variance.

Large decision trees are probably used most commonly as autonomous base learners because small random changes produce large changes in the tree (large variance) and large trees reproduce the training data well (small bias). Another model such as a support vector machine might be better than a single decision tree for the problem at hand, but the model that is best by itself might not have the properties of high variance and low bias needed for an ensemble of autonomous models to work. In one paper on a type of ensemble called bagging, Breiman (1996) shows that stepwise regression can work in the same way that decision trees work. Over time the ensemble approach has been closely associated with decision trees so that most ensemble results employ trees even though other predictive approaches would also likely work, perhaps just as well.

Ensembles in Discovery

There is more recognition and use of multiple decision trees even in the areas of discovery and interpretation (where the traditional simplicity of single decision trees favors them). Ensembles help counter the tendency for small changes in the data set to produce substantial changes in the resulting tree. So, although it is quicker and easier for purposes of exploration to identify relationships with a single tree, this advantage is offset by the uncertainty of the potentially unstable result. It is perhaps too easy to move down the wrong path in the early stages of data analysis that relies solely on single decision trees for data exploration.

One role of exploration is to examine the vagaries of a seemingly inconsequential number of observations in a study data set. Sometimes, removing observations that seem inconsequential can result in different variables being important in the tree and different leaf-segments being identified.

Even in the preliminary stages of data analysis, more reliable discoveries of important variables, rules, and segments emerge from the use of multiple trees. The improved results offset the additional time and effort that multi-tree approaches require. Advances in visualization make it increasingly practical to employ these multi-tree approaches even for preliminary examination and exploration tasks.

New algorithms are beginning to emerge that combine the multi-tree approach of ensembles together with the simplicity of interpretation that single-tree approaches offer. One such approach is described by De Comité et. al. (2003).

The Evolution of Ensemble Methods

Ensembles began to emerge in the 1980s as computer power and access began to foster the development of more computationally intense methods of tree growth.

Cross-Validation

The decision tree approach that was developed by Breiman, et. al. (1984) introduced a method of validation, frequently referred to as cross-validation, which employs multiple decision trees. Cross-validation is different from the split-sample or hold-out method. In the split-sample method, only a single subset (the validation data set) is used to estimate the generalized error rate. Cross-validation uses several subsets of the available data to estimate generalized error rates. The resulting estimates are often used when choosing a model.

In the classical split-sample or hold-out method of validation, the data set that is used to train the statistical model is first sampled so that a smaller representative sample of the host data can be used at the end of the training step for test and validation purposes. One potential problem with classical validation is that the creation of this validation sample takes away instances of data that could otherwise be used in training the statistical model.

The cross-validation technique developed by Breiman et. al. (1984) addresses the issue of data loss through validation sample construction by using all the original data for training purposes. The original training data is re-sampled a number of times and each instance of re-sampling is then used as a contributor to the construction of a synthetic validation data set. In this way, a synthetic hold-out sample is created that can serve as a proxy validation data set, and all training records are used in statistical training.

Cross-validation is central to Breiman's grow-and-prune approach to constructing a tree that is both reliable and accurate. Here, the parent decision tree is allowed to grow large because cross-validation, performed through the use of the proxy validation data set, is then used to trim bottom branches that have marginal validity. The trimming is taken care of by computing a cost-complexity measure that tries to produce a "right-sized" or "honest" tree. This approach hinges on the understanding that an overly-complex tree will fit the training data very well but will not generalize to new data. Cost complexity is introduced by Breiman et al. (1984).

The Breiman style of cross-validation works as follows: The parent data set is partitioned into groups called folds. Typically, 10 folds are used (which is called 10-fold cross-validation). Nine of the partitions are used as a new cross-validation training data set. The 10% (1 out of 10 partitions) of the data that was held back from the cross-validation training data set is used as an independent test sample for the test decision tree. This data is run through the test decision tree and the classification error rate for that data is computed. This error rate is stored as the independent test error rate for the first (1 of 10) test decision tree.

A different set of nine partitions is now collected into a cross-validation training data set. The partition held back this time is different from the partition held back for the first test decision tree. A second test decision tree is built and its classification error rate is computed. This process is repeated 10 times, building 10 separate test decision trees. The cost-complexity calculation is used to determine the stopping rule for tree growth. In each case, 90% of the data is used to build a test decision tree, and 10% is held back for independent testing. A different 10% is held back for each test decision tree. Once the 10 test decision trees have been built, their classification error rates (which are a function of the decision tree size) are averaged. This averaged error rate for a decision tree is known as the cross-validation cost. This cross-validation cost can then be used as if a fresh hold-out sample had been used and so a measure of the generable applicability of the tree model can be produced.

Bootstrapping

In 1979, Bradley Efron introduced the bootstrap method for estimating the distribution of an arbitrary statistic. Some statistics such as the average have formulas for estimating the variance and other features of the distribution. Other statistics such as the median have no neat formula for the variance. The bootstrap is a neat numerical method for computing the variance of these statistics.

The bootstrap process begins with the original data set of observations and forms a bootstrap sample by repeatedly selecting an observation from the original data set at random. The same observation can be selected many times, but more than likely, different observations will be selected. Because the same observation can be selected more than once, this is called sampling with replacement.

A bootstrap sample usually contains as many observations as the original training data, so if you began with 1,000 observations, you would have 1,000 observations in the bootstrap sample. Many of the observations are duplicates. Bootstrapping consists of constructing many bootstrap samples (for example, 50, 100, or even 2,000, depending on what you do). The statistic of interest is evaluated on each sample and saved in a pool. Displaying the pool of computed statistics indicates how precise the statistic computed with the original data is.

Bagging

In 1996, Leo Breiman introduced the bagging method for generating multiple decision trees and combining their predictions into a single prediction. Bagging stands for bootstrap aggregating. The bagging process is shown in Figure 5.17.

Bagging uses bootstrap samples to inject random changes into base learners. In a bootstrap sample, the same observation can occur more than once. The number of observations in a bootstrap sample typically equals the number of observations in the original data set.

Sampling without replacement refers to a sample in which an observation can occur at most once. The number of observations in the sample is less than the number in the original data set. When data sizes are small, every observation is important to the base learner. As the size of the training data increases, the average contribution of an individual observation to the prediction accuracy of the model decreases. Accordingly, sampling with replacement is appropriate for small data sizes in order to preserve the number of observations. However, for larger data sets, sampling with replacement is not necessary and can be less effective than sampling without replacement (because bootstrap samples are more correlated than samples without replacement).

Figure 5.17: A High-Level Illustration of Bagging

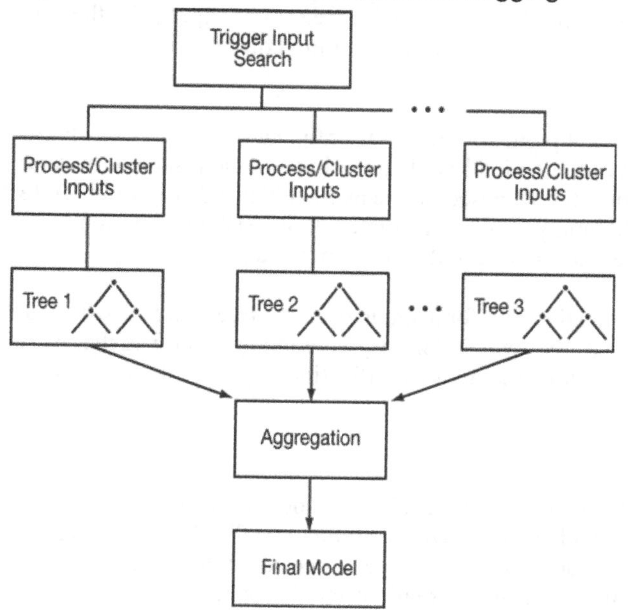

Because of the sometimes erratic results produced by injecting random changes in base learners, bagging is now rarely used. It is useful to see the bagging approach as an exploratory step in the development of Random Forests. This helps us understand that Random Forests incorporate notions from Bagging and improve upon them. This background helps explain the leading position of random forests as the standard ensembles of autonomous models today.

Holographic Decision Trees

The paths of artificial intelligence, cognitive science, and data mining have been interwoven from the original development of the first decision trees. Their relationship has stayed in place and is stronger now than it has ever been. No place is this more obvious than in the development of what has become one of the standard forms of multiple decision trees—random forests.

Some of the first published results on random forests were originally reported by Amit and Geman (1997). The authors were interested in the effects of randomization on the construction of decision trees and adopted the idea of choosing a random sample of predictors from the collection of inputs at each node of a decision tree. They referred to their work as the construction of a kind of holographic decision tree. The resulting decision tree is holographic in the sense that each node has the possibility of reflecting a different facet of the predictive space that is contained in the training data. This was accomplished by taking a random sample of available predictors or inputs available at each node.

At each node, after the sample of predictors was taken, an estimate of the best predictor was made using a random sample of data points. After producing *n* decision trees (T1, T2, …, T*n*) the authors picked a predictive structure designed to maximize the average terminal distribution in the resulting decision tree.

At this point, they chose a random sample of predictors from the entire candidate data set. From this sample, they developed an estimate of the optimal predictor using a random sample of data points. They then developed a maximum estimate based on the average (leaf) distribution. To test their approach, the authors used 223,000 binary images of isolated digits written by more than 2,000 writers. They used 100,000 images for training and 50,000 for testing.

What they found was that the best classification rate for a single decision tree was 5%, whereas the best classification rate of multiple decision trees was 91%. They convincingly showed that by aggregating decision trees, the success of automatic digit classification improves dramatically.

Random Forests

Leo Breiman continued his earlier work on bagging and evolved a random forest approach that came out at about the same time as Amit and Geman's work (Breiman, 1999). Like bagging and the Amit and Geman precursor, a random forest is an ensemble of decision trees. Multiple decision trees are grown, each based on a random sample of observations that is taken from the original, host training data. Because multiple trees are produced from one host training data set, the result is called a forest.

The general computational approach, as described by Breiman (2001), is as follows:

- Denote the training data as N and the number of inputs as M.
- Take a bootstrap sample (i.e. sample with replacement) from N, designated as n (referred to as the bagged data).
- For each n sample, set aside the remaining cases for error estimation (referred to as the out-of-bag data.
- At each stage of the tree growth, randomly select a smaller number of inputs, designated as m.
- For each node to be split, take the best split based on the m variables that have been selected; i.e., select the split with the lowest error. Compute the error by running the prediction for n and comparing the result to the actual values contained in the sample that is set aside.
- Grow each tree fully and do not prune them.

For prediction, a new sample is run through the multiple trees. Each observation is assigned the label of the training sample in the terminal node that it ends up in. This procedure is iterated over all trees in the ensemble. The modal vote of all trees is reported as the composite estimate.

The HPFOREST Procedure

The multiple trees can be spawned as separate, independent processes that can be implemented as separate, parallel process streams. The organization of this approach is well adapted to high-performance computing because each tree can be assigned to a separate computing processor. The high-performance adaptation of the forest algorithm is implemented as a SAS procedure: HPFOREST (the prefix HP refers to "high performance").

The HPFOREST process is illustrated in Figure 5.18. The process begins by taking a random sample (without replacement) of the available rows in the training data. This sample serves as the source for growing one of the decision trees that are grown as part of the forest of trees that form this solution. In the terminology of this method, the random sample is set aside to be used as the training data for the current tree that is being trained; i.e. it is bagged. A portion of the random sample is set aside as a local, test sample, called the out-of-bag (OOB) sample. Different trees have different out-of-bag samples because they use different random samples for training.

Figure 5.18: HPFOREST Process Flow Description

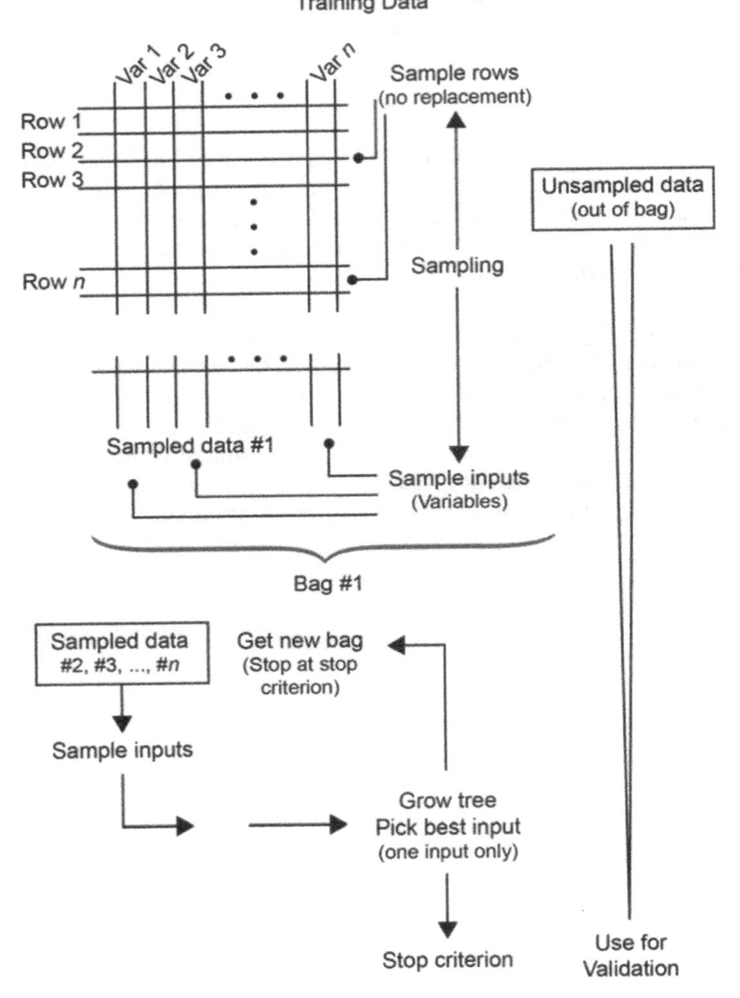

The quality of a forest model is typically measured with a goodness-of-fit statistic such as the misclassification rate or the average square error. A fit statistic is often a simple average of a function of the actual and predicted target value of an observation. When the average is taken over the training observations, the fit statistic gives an overly optimistic impression of how the model will perform on future data. The OOB samples provide a way to get a more pessimistic impression. For a specific observation, compute its prediction without using trees that use the observation for training. If the forest contains 100 trees and the observation is OOB in ten trees, then compute the prediction using only those ten trees. Call this the OOB-prediction. The OOB-predictions for two observations are generally formed from different trees and a different number of trees because the two observations are generally in the OOB samples of different trees. An OOB fit statistic is the average over the training observations of a function of the actual target value and the OOB-prediction. The OOB fit statistic gives a slightly pessimistic impression of how the forest will perform on future data. This built-in validation feature of the HPFOREST procedure is a strong point of the process.

As in bagging, decision trees are grown independently (in parallel). As Breiman points out in his discussion of forests, the randomness makes the variable selection less greedy (i.e., less likely to over fit) thus mitigating the need for pruning. Because many decision trees are grown, the expectation is that in the long run, the better variables are more likely to be selected. The development of a forest solution can require the development of hundreds, if not thousands of individual decision trees

At each node of the developing decision tree, a subset of inputs is selected at random out of the total number of inputs that are available. The number of inputs competing to be used for splitting a node is therefore less than the total number of inputs present. The HPFOREST procedure selects the competing inputs that have the highest association with the target variable, and then finds the best splitting rule with the selected input. The alternative method is to find the best splitting rule for each competing input and use the one with the best worth. Because this method cultivates trees based on pre-viewing and selecting inputs with high-worth splits, it has a built-in variable selection bias that can weaken its generality.

Stopping Criteria

The original forest algorithms grow trees until each leaf has a single observation. Training with 1,000 observations would produce a tree with 1,000 leaves. Trees in a forest are no different than single trees with respect to overfitting the training data. A tree that grows to singletons is less accurate on future data than a subtree obtained by pruning nodes. The trouble is in choosing which nodes to prune. Inventors of forests say that the solution is elusive. They propose growing many large trees and letting the random errors from overfitting cancel when the predictions from different trees are averaged. It works. Although a forest of smartly pruned trees might perform better, the improvement is presumably small. It is hard to know for sure because it is hard to prune trees smartly.

Computational performance is another matter. Training with a million observations would produce a tree with a million leaves and take at least 1,000 times longer to train than a tree with 1,000 leaves (1,000 times1,000 equals 1 million). Scoring a new observation would take about 10 times longer with a million leaves than with 1,000 leaves. Limiting the growth of the tree might be wise for computational reasons.

As with a single decision tree, stopping criteria can be set up to stop tree growth at a given depth, node size, or alpha level. The HPFOREST procedure specifies a threshold p-value for the significance level of a test of association of an input variable with the target. If no input has an association above the threshold, then the node is not split.

The HPFOREST procedure has an option for specifying the number of trees in the forest. Typically, the fit statistics improve as the number of trees that are grown increases until diminishing returns result whereby new trees fail to improve model performance. Finding the number of trees to grow varies in each new case, so it is subject to trial and error.

Scoring

As with other ensembles, scoring is a function of creating the summed average vote of all the constituent trees in a forest for each instance that is to be scored. To score a new data set, pass each row of the new data through each decision tree in the random forest and record the predicted value that is given by each decision tree. To aggregate the results, either compute an average of all the scores (for a continuous or categorical target) or determine the most likely class value through a majority rule from the classifications that are produced by the decision trees (for a categorical target).

Advantages of Random Forests as Predictive Models

As previously indicated, the formation of the OOB data provides a validation measure. The OOB data is the residual data left over through the creation of any one decision tree. These samples are formed without replacement. The OOB data therefore has a slightly different profile than the bagged data that is used for training. It turns out that this characteristic of the OOB data makes it particularly appropriate as a measure of how the forest ensemble will likely perform on new data. So forest accuracy measures appear to hold up exceptionally well as the models are generalized to new data sources.

The successive sampling of training data followed by sampling inputs as potential sources of split branches also appears to offer many benefits. Each tree in the ensemble provides a finely fit sub-region of a total solution which, in aggregate, produces more accurate results than a single model approach.

Breiman's research with random forests showed that in many cases, random forests are more accurate than boosting approaches. He found that the random forest approach could handle hundreds and thousands of input variables with no degeneration in accuracy. In his latest work, he presented many unique ways to present the results of random forests and proposed methods for using random forests to do a form of cluster analysis (Breiman 2001).

Boosting Multiple Trees

Boosting refers to combining simple, inaccurate modeling algorithms into a more accurate ensemble. Common boosting methods grow trees in a sequence with each successive tree focusing on improving the weaknesses of the trees already grown. As previously discussed, growing trees autonomously in parallel and averaging their predictions improves accuracy. When using a boosting approach trees are grown sequentially, rather than autonomously.

Averaging autonomous ensembles of trees produces superior results by making the variance of the ensemble smaller than that of a single tree. Boosting works by a different mechanism. In fact, boosting can improve accuracy without reducing the variance. Building an ensemble of autonomous trees requires inserting random changes into the training data or into the algorithm. Boosting does not require making random changes (although it can help).

The boosting approach was developed by Schapire (1989). AdaBoost by Freund and Schapire (1996), arcing by Breiman (1998), and gradient boosting by Friedman (2001) (which is discussed below), extends the original development. Boosting has been incorporated in many software solutions, including the development of the C4.5 and C5 toolkits by Quinlan (Quinlan, 1993).The TREEBOOST procedure in SAS Enterprise Miner incorporates the boosting approach.

To define a tree ensemble, you must specify how the trees differ and how the predictions combine. Common boosting methods create different trees by changing the training data of successive trees. AdaBoost changes the weight of each observation. Arcing changes the sample. Gradient boosting changes the target values. In each case, the new tree focuses on observations that the previous trees do not predict well. This inclines all of these boosting approaches to a successively refined composite result.

Stochastic gradient boosting (sometimes called gradient boosting), developed by Friedman (Friedman, 2001), changes the target values and also changes the training data (by sampling without replacement). The stochastic version was introduced shortly after an original non-stochastic version (Friedman, 1999a) and improved accuracy and performance. It is now the version regularly in use.

Gradient boosting differs from earlier boosting methods, such as arcing, by ensuring that every observation in the training sample has an equal chance of inclusion. (In this way, it employs the approach that is routinely employed with the development of multiple, autonomous ensembles.) By contrast, arcing starts with bagging and modifies the sampling to favor observations that are predicted poorly by the previous trees. In a role reversal, arcing takes an approach from multiple, autonomous tree ensembles and grows large trees. In contract, gradient boosting uses small trees.

AdaBoost combines the predictions of the individual trees by taking a weighted sum of the votes for each class and using the weighted majority class as the prediction of the ensemble. Arcing takes an unweighted vote. Gradient boosting computes posterior probabilities for the ensemble and uses the class with the largest probability as the prediction.

All these approaches have demonstrated superior results on different sets of data. There are a number of theories about why they work. Practitioners tend to explain their results based on the approach that works for them (and that they are familiar with).

AdaBoost

The AdaBoost algorithm was developed by Freund and Schapire (1996). AdaBoost reweights individual observations before training the next tree. The weights influence the tree algorithm. AdaBoost iteratively adjusts the case weights and adapts the tree algorithm to them. The authors view this as an adaptive boosting algorithm (which explains the origin of the "AdaBoost" name).

AdaBoost is a form of boosting that builds an initial model from the training data set. This first pass through the training data results in a standard model. From this model, some records are correctly classified by the decision algorithm, and some records are misclassified. A formula converts the proportion of misclassified to a number W(1). The misclassified observations are given a weight proportional to W(1). The correctly classified ones are given a weight proportional to 1/W(1).

The second tree is trained with the weights derived from W(1). From this second model, some records are correctly classified and some records are misclassified. A formula converts the proportion misclassified in the second tree to a number W(2). The weight of an observation misclassified by the second tree is multiplied by W(2). The weight of a correctly classified observation is multiplied by 1/W(2). The weights are then rescaled to sum to one. The third tree uses these new weights.

The building of the initial model, followed by boosting, is repeated until the incrementally generated model performs at the level of a random guess. This indicates that forming additional boosted samples is not likely to contribute worthwhile results. At this point, there is a panel of models. This panel is used to make a decision on new data by combining the expertise of each model so that the more accurate experts carry more weight.

Gradient Boosting

The SAS TREEBOOST procedure implements stochastic gradient boosting (Friedman 2001).

Gradient boosting changes the data for training successive trees changes in two ways: The target is the residual of the original target from the current prediction, and the training data for one tree is a sample without replacement of the available data.

The residual that is calculated in the series is defined in terms of the derivatives of a loss function. For an interval target, the derivate of squared error loss results in a residual that is simply the target value minus the predicted value. Because extreme target values can overly influence the loss function, Friedman proposed the Huber-M loss function as an alternative that reduces the weight of observations with extreme target values. With categorical targets squared error loss can also be calculated. In this case, the residual is either the current posterior probability of the event class, or 1 minus this posterior probability. Boosting is defined for binary, nominal, and interval targets.

Comparing Gradient Boosting and Forests

Tree-boosting machines and decision tree forests are predictive models that combine the predictions of many decision trees. They are generally better predictors than a single tree, and retain most of the appeal of the traditional form of decision trees: Categorical inputs and missing values are handled seamlessly (without creating separate binary variables for separate categories) and binary searches of interval inputs are not unduly influenced by extreme values of the inputs.

A disadvantage of boosting machines and forests arises from the methodology of using many—often partial—trees to form a prediction. The multiple trees are grown and combined so many times that the typical transparency of the decision tree display is obscured, and so is not easily interpretable. In most cases, the tree structure is so obscured that the only available information is the importance rank of the input variables.

Gradient boosting produces final scores in various fashions. For a categorical target, the final score is produced by combining the predictions of the individual trees. Scores are formed either by vote or by computing the average estimated class probabilities of each category (often referred to as the posterior probabilities). *Voting* means that the boosting machine or forest predicts the target to have the category predicted by majority vote of the combined individual trees. Using voting, the overall posterior probability of a target category is the proportion of votes assigned to that category if the split were made on a given input. Voting sometimes produces different predictions than averaging probabilities. Neither method is generally better than the other. For an interval target, the overall prediction is the average of the predictions of the trees.

The difference between boosting and forests is that the training data for an individual tree in a boosting machine depends on the predictions of the trees already trained. Trees in a forest are formed from a series of autonomous, independent samples. Trees in a boosting machine are generally small; trees in a forest are generally large. A small tree is not very accurate on most data sets and sometimes not useful at all. In machine-learning terminology, a small predictive tree is a *weak* model, one that predicts marginally more accurately than random guessing. A *strong* model is one that, given enough data, can reduce the prediction error to the Bayes limit, the limiting error rate inherent in the data. *Boosting* converts a weak algorithm into a strong one. But if a tree is too weak, if its predictions are not sufficiently better than chance, boosting will not help.

Leo Breiman introduced random forests (2001) in an attempt to extend the ideas of others and produce a global approach. Trees in his forests differ because randomness is injected at two places: the training data is a random sample of the available data, and the candidate splitting variables in a node are a random sample of the available inputs. The role of randomness is to reduce the correlation between trees. Breiman noticed that the prediction error for an observation contains a variance component, and the variance over the forest is smaller if the correlations between tree predictions of the observation are smaller.

Adding trees to a forest will not over fit the data. Tim Kam Ho (1995) proposed averaging a forest of large trees as an alternative to the elusive task of finding the right size of a single tree. The generalization error of a single tree first improves as the size of the tree increases, and then deteriorates as over fitting increases with larger tree sizes. Ho and others who followed grew large trees, believing that the errors from over fitting a single tree will cancel when averaged over many trees. However, Breiman's theorem (2001) indicates that the generalization error converges as the number of trees increases (there is no indication that the limiting error is as good as it would be if the individual trees were right-sized). The benefit of right-sizing might be small and might not justify the extra effort involved in determining which size of tree is best.

Adding trees to a boosting machine can eventually over fit the data. This is because the trees—which are sampled from the same population—are not independent. In general, the prediction error of an observation decomposes into bias and variance components. In a typical boosting machine, the resulting trees are simple and therefore have a large bias. But the bias of the entire boosting machine generally decreases as more trees are added, because the newer trees are focused on observations that the previous trees did not fit well. However, the variance might not always decrease. Consequently, the prediction error can begin to rise as the number of trees that are included also rises. Forests have the opposite effect: The bias stays the same as more trees are added, and the variance decreases. Forests generally approach their level of accuracy with fewer trees than boosting. However, boosting often eventually settles to a lower error rate than forests. Both methods have respective strengths and weaknesses.

Chapter 6: The Integration of Decision Trees with Other Data Mining Approaches

Introduction

So far most of the discussion on decision trees has been in the area of their direct utility as descriptive and predictive tools. This chapter shows how decision trees can also be used to provide software support when integrating decision trees with other data mining techniques. The following topics are addressed in this chapter:

- stratified regression
- decision trees in forecasting applications
- decision trees in variable selection, interaction detection
- decision trees in analytical model development

- decision trees in rule induction
- decision trees for multivariate analysis

Recall that the original use of decision trees was as a complement or alternative to regression. As decision trees have developed, their abilities as a complement to, as well as a substitute for, other data mining techniques have increased.

Decision Trees in Stratified Regression

Stratified regression is one of the oldest applications of decision trees. As illustrated in Figure 1.3 in Chapter 1, the goal of stratified regression is to divide the main data set into subgroups so that different regression equations fit into each of the subgroups. This is especially appropriate when the differences in the subgroups are so profound that it is simpler and more effective to determine the specialized shape of the regression equation in each of the subgroups.

A formal definition and extensive explanation of stratified regression is provided in Neville's article on stratified regression (1999). Boston housing data is used as a basis for the examples in this article, and the StatLib repository (http://lib.stat.cmu.edu/) is the source of the original data. The following list describes the variables that are included in the data set. The dependent variable is **Lmv**.

Name	**Description**
Crim	Per capita crime rate by town.
Zn	Proportion of a town's residential land zoned for lots greater than 25,000 square feet.
Indus	Proportion of nonretail business acres per town.
Chas	Charles River dummy variable.
Noxsq	Nitrogen oxide concentration (parts per hundred million) squared.
RmSq	Average number of rooms squared.
Age	Proportion of owner-occupied units built prior to 1940.
Dis	Logarithm of the weighted distances to five unemployment centers in the Boston region.
Rad	Logarithm of index of accessibility to radial highways.
Tax	Full-value property tax rate.
Ptratio	Pupil-teacher ratio by town.
B	(Bk – 0.63) squared, where Bk is the proportion of blacks.
Lstat	Logarithm of the proportion of the population that is lower status.
Lmv	Logarithm of the median value of owner-occupied homes.

Neville's article explains that there are several ways to form strata based on decision trees. From these strata, regression models were produced that were better than regression models based on fitting the entire data set as one block of data. Figure 6.1 illustrates the decision tree branch with one of the possible strata that could be used to improve the regression results. The branch split on nitrogen oxide concentration (parts per hundred million) squared (**Noxsq**).

Figure 6.1: Illustration of Decision Tree Identifying Strata

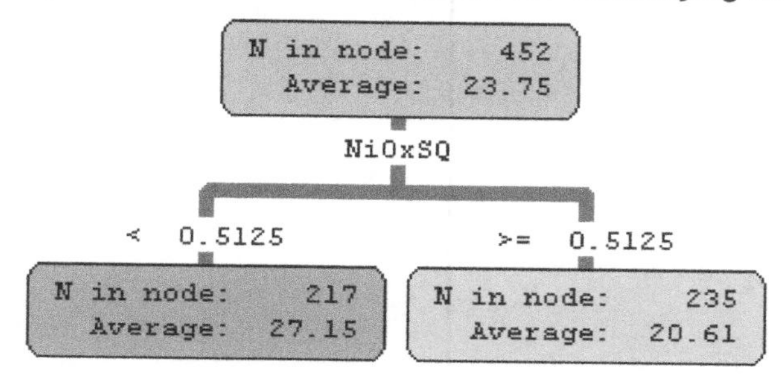

Tax and ? (missing) are also strata that could improve the regression results.

To implement stratified regression, it is necessary to partition the data. Alternatively, you can create an effect variable that partitions the data numerically. In the case of strata formed by nitrogen oxide, you should form two regression equations. One regression equation will fit the data on the left (217 observations), and one regression equation will fit the data on the right (235 observations).

Time-Ordered Data

Traditionally, decision trees have been used to analyze cross-sectional data such as survey data. A cross-sectional data set contains measurements of a variety of observations at a given point in time, as shown in Figure 6.2.

Figure 6.2: Illustration of the Form of Data Captured from One Point in Time

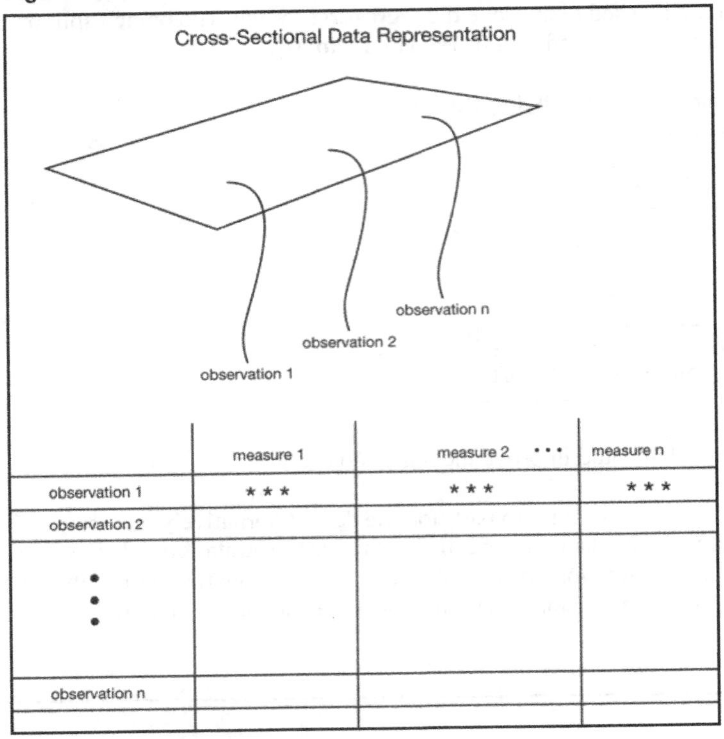

Cross-Sectional Data Representation

observation n

observation 2

observation 1

	measure 1	measure 2 • • •	measure n
observation 1	* * *	* * *	* * *
observation 2			
• • •			
observation n			

By contrast, time series data contains measurements of a variety of observations at various time intervals (e.g., stock prices), as shown in Figure 6.3.

Figure 6.3: Illustration of the Form of Data Captured from a Time Series

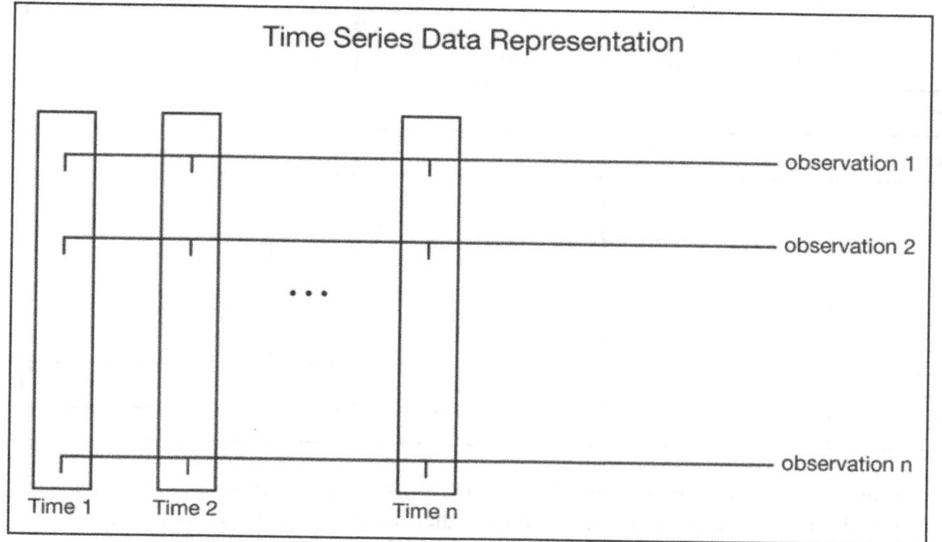

Decision Trees in Forecasting Applications

As analysts have gained more experience with time-ordered data, they have discovered that techniques that are commonly used to analyze cross-sectional data can be adapted to analyze time series data. Thus, regression techniques are used in the analysis of both cross-sectional and time series data. Similarly, decision trees can be used to analyze both cross-sectional and time series data.

Figure 6.4 illustrates a typical time series in cross-sectional form, where **m** stands for measure and **t** stands for time.

Figure 6.4: Illustration of Reworking Time Series Data into Cross-Sectional Form

	M_1T_1 M_2T_1 ... M_nT_1	M_1T_2 M_2T_2 ... M_nT_2	• • •	M_nT_1 M_nT_2 ... M_nT_n
observation 1				
observation 2				
• • •				
observation n				

In Figure 6.5, a time series shows the rise and fall of lynx traps in any given year as the lynx population rises and falls according to the operation of other factors in ecology, such as food supply, disease, and predators. The lynx population hits a peak in the years of 1828, 1866, and 1904.

Figure 6.5: Illustration of a Time Series

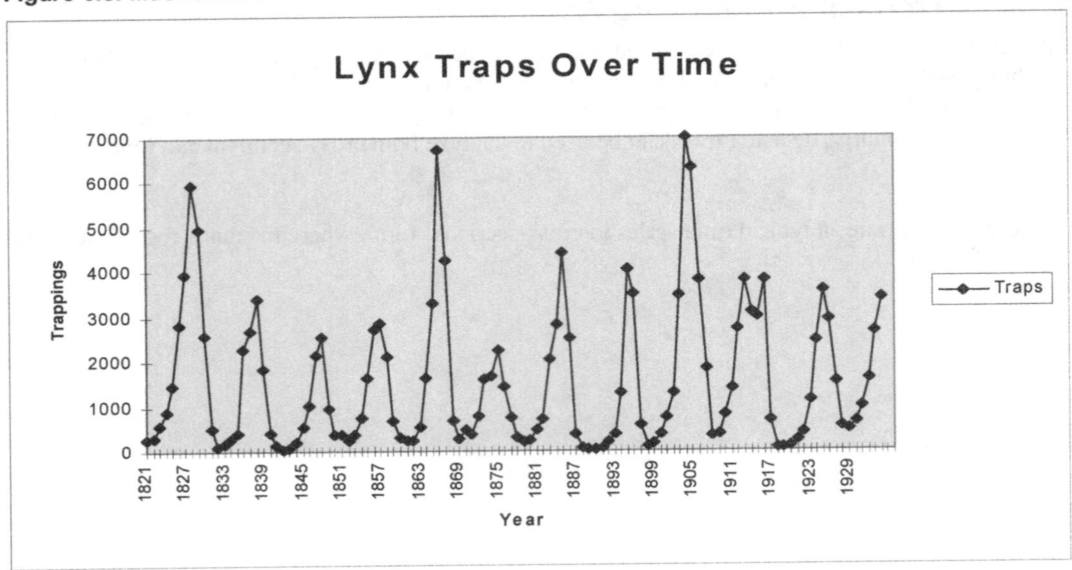

A portion of the data underlying these results is shown in the following list.

Year	Traps
1821	269
1822	321
1823	585
1824	871
1825	1475
1826	2821
1827	3928
1918	81
1919	80
1920	108
1921	229
1922	399
1923	1132
1924	2432
1925	3574
1926	2935
1927	1537
1928	529
1929	485

Figure 6.6 shows a decision tree that reproduces the results of the graph in Figure 6.5. The peaks in the graph are captured in the intervals of 1827–1830, 1864–1867, and 1903–1906 (the lowest interval of <= 1826 and the highest interval of >= 1907 are not shown in the decision tree).

Figure 6.6: Illustration of a Decision Tree of the Lynx Time Series Data

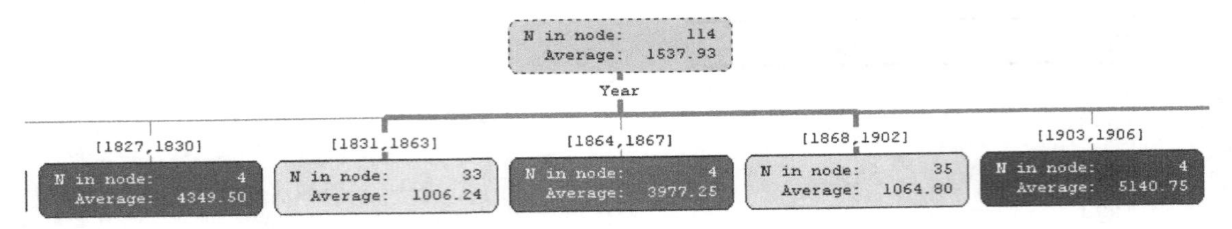

In many situations, a decision tree handles time series data in a straightforward way. This is shown in the classic study of lynx traps. In many applied situations, some reworking of the data might be necessary. For example, in direct marketing there is a need to derive customer measures for recency, frequency, and monetary value. These measures come from transaction data based on purchase interactions. For recency, you can sum all transactions and create measures, such as last purchase date. For frequency, you can count the number of monthly purchases. You can use total or average purchases for monetary value.

Banks like to distinguish card account holders by their purchase habits. Do the card holders use the card a lot, and pay down the outstanding balance on a monthly basis? Do they consistently maintain an outstanding balance? Has a customer moved from one mode of payment to another? If so, why? Distinguishing card account holders in these ways means that fields must be created to measure these characteristics on a monthly basis. Furthermore, the characteristics need to be stored on a month-to-month basis and, if a characteristic changes, an indicator needs to be set. This results in the creation of a new field of data.

Decision Trees in Variable Selection

In the following banking data set, the goal is to determine the attributes of online transactions.

Figure 6.7: Bank Data Online Transactions

Name	Role	Report	Level	.	.	L	U	.	Label
DDA	Input	No	Binary		No	.	.	N	Checking Account
DDABAL	Input	No	Interval		No	.	.	N	Checking Balance
DEP	Input	No	Nominal		No	.	.	N	Checking Deposits
DEPAMT	Input	No	Interval		No	.	.	N	Amount Deposited
DIRDEP	Input	No	Binary		No	.	.	N	Direct Deposit
HMOWN	Input	No	Binary		No	.	.	N	Owns Home
HMVAL	Input	No	Interval		No	.	.	N	Home Value
ILS	Input	No	Binary		No	.	.	N	Installment Loan
ILSBAL	Input	No	Interval		No	.	.	N	Loan Balance
INAREA	Input	No	Binary		No	.	.	N	Local Address
INCOME	Input	No	Interval		No	.	.	N	Income
INV	Input	No	Binary		No	.	.	N	Investment
INVBAL	Input	No	Interval		No	.	.	N	Investment Balance
IRA	Input	No	Binary		No	.	.	N	Retirement Account
IRABAL	Input	No	Interval		No	.	.	N	IRA Balance
LOC	Input	No	Binary		No	.	.	N	Line of Credit
LOCBAL	Input	No	Interval		No	.	.	N	Line of Credit Balance
LORES	Input	No	Interval		No	.	.	N	Length of Residence
MM	Input	No	Binary		No	.	.	N	Money Market
MMBAL	Input	No	Interval		No	.	.	N	Money Market Balance
MMCRED	Input	No	Nominal		No	.	.	N	Money Market Credits
MOVED	Input	No	Binary		No	.	.	N	Recent Address Change
MTG	Input	No	Binary		No	.	.	N	Mortgage
MTGBAL	Input	No	Interval		No	.	.	N	Mortgage Balance
NSF	Input	No	Binary		No	.	.	N	Number Insufficient Fund
NSFAMT	Input	No	Interval		No	.	.	N	Amount NSF
PHONE	Input	No	Nominal		No	.	.	N	Number Telephone Banking
POS	Input	No	Interval		No	.	.	N	Number Point of Sale
POSAMT	Input	No	Interval		No	.	.	N	Amount Point of Sale
RES	Input	No	Nominal		No	.	.	C	Area Classification
SAV	Input	No	Binary		No	.	.	N	Saving Account
SAVBAL	Input	No	Interval		No	.	.	N	Saving Balance
SDB	Input	No	Binary		No	.	.	N	Safety Deposit Box
TELLER	Rejected	No	Interval		No	.	.	N	Teller Visits
online	Target	No	Interval		No	.	.	N	number of online transactions

A traditional variable-importance approach looks at the zero-order correlations between all possible inputs and the target. This approach produces the following chart:

Input	Correlation
DEPAMT	0.3597
POSAMT	0.21106
POS	0.18395
ACCTAGE	0.13256
DDABAL	0.11238
ATMAMT	0.10031
NSFAMT	0.08569
ILSBAL	0.08247
LOCBAL	0.07963
HMVAL	0.06979
AGE	0.0289
CCBAL	0.02787
MTGBAL	0.01641
INCOME	0.01173
CRSCORE	0.01077
LORES	0.00613
SAVBAL	0.00116
IRABAL	-0.01121
INVBAL	-0.02104
MMBAL	-0.03722

The strength of the correlation is shown in Figure 6.8.

Figure 6.8: Correlation between Inputs and Target

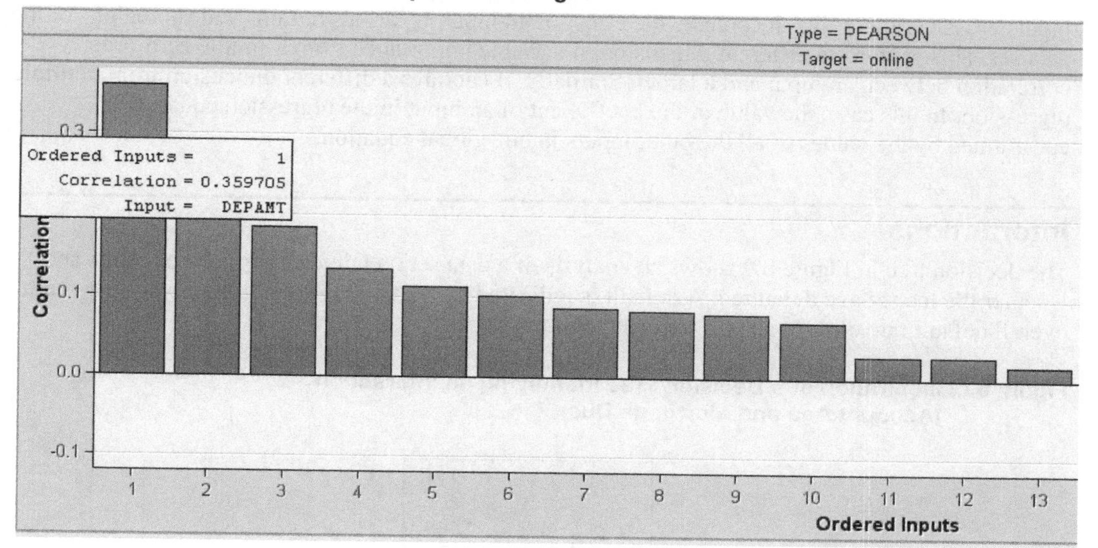

Decision Tree Results

The decision tree results tell a different story. In the decision tree, variable importance is calculated as the sum of the worth statistics for an input across all the split nodes of the decision tree. If an input is an important splitting criterion in many levels of the decision tree, then its importance grows as a result. Inputs that do not appear in any splits have zero importance.

DEPAMT	Amount Deposited	1
DEP	Checking Deposits	0.34
CCBAL	Credit Card Balance	0.19
ACCTAGE	Age of Oldest Account	0.18
DDABAL	Checking Balance	0.09
ATMAMT	ATM Withdrawal Amount	0.09
LOC	Line of Credit	0.06
POSAMT	Amount Point of Sale	0.05
MTG	Mortgage	0.05
DIRDEP	Direct Deposit	0.05
AGE	Age	0.05
PHONE	Number Telephone Banking	0.05
ATM	ATM	0.04

Because the decision tree method of calculating variable importance incorporates the effect of an input across various splits, it captures the effect of an input in various regions and subregions of the data set. This captures a different dimension of variable importance from a simple zero-order correlation between an input and a target. Similarly, it captures a different dimension from multiple regression. In this case, the value of the coefficient of an input in the regression equation is constrained by the values of all the other inputs in one global equation.

Interactions

The decision tree in Figure 6.9 shows an analysis of a data set on home equity loan histories and whether the loans have defaulted. A default is indicated by a Bad=0 field in the analysis. The high overall default rate of 80% is used to illustrate interactions.

Figure 6.9: Illustration of a Decision Tree Identifying an Interaction (Account Age and Mortgage Due)

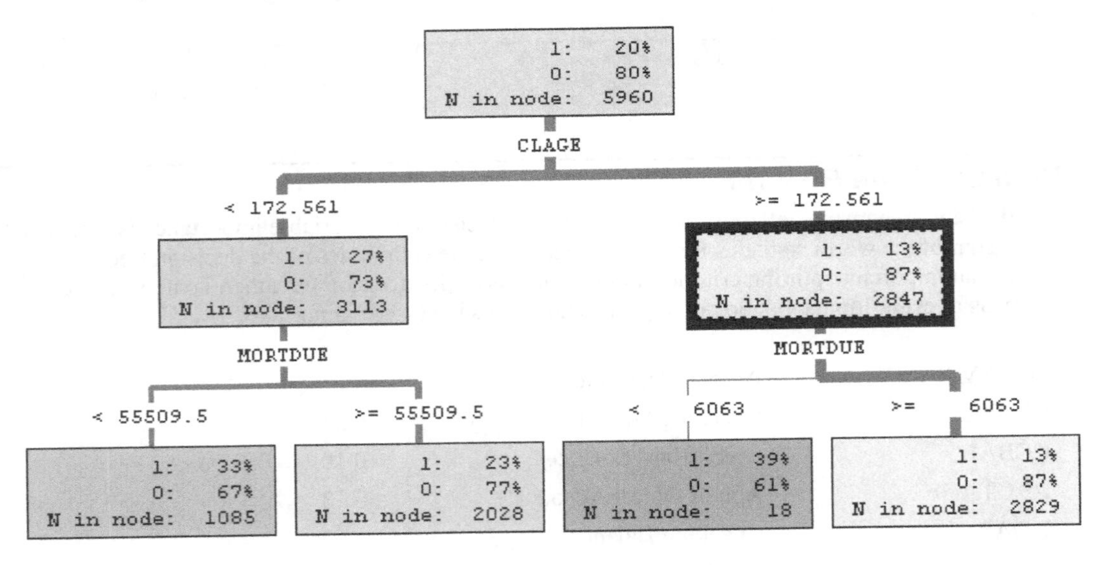

Defaults increase in parallel with how old the outstanding credit line is. The decision tree shows 73% default for credit lines that are up to 172 days old, and 87% default for credit lines that are older than 172 days.

The lower level of the decision tree shows an interaction between credit-line age and the amount of mortgage due. When the age is less than 172 days, the greater amount of mortgage due increases the default rate by 10%—from 67% to 77%. When the age is equal to or greater than 172 days, the greater amount of mortgage due increases the default rate from 61% to 87%—a difference of 26%. This is more than twice the increase in the younger credit lines.

This suggests that it could be useful to construct an interaction term that combines age and mortgage due (clage*mortdue) when building the regression model to predict default. In this example, the direction of the interaction on the left and right of the decision tree is the same. On the lowest level of the decision tree on both sides, the decision tree is formed by the same branch (created by using **MORTDUE** as a branch split input). This decision tree and its interactions can be considered symmetrical.

It is possible to have asymmetric interactions in decision trees. The subtree formed by **MORTDUE** might produce different directions of the interaction depending on the side of the decision tree. This is a reversal; envision that the left side of the decision tree contains leaves with 77% and 67% defaults, rather than 67% and 77% as seen in Table 6.1.

Another form of asymmetry is when there is a different partitioning field on the right side of the decision tree compared to what is on the left side.

Cross-Contributions of Decision Trees and Other Approaches

Table 6.1 describes the various impacts between data mining methods.

Table 6.1: Data Mining Methods Cross-Impact Matrix

Data Mining Methods Cross Impact Matrix					
	Association	Clustering	Regression	Decision Trees	Neural Networks
Association				Create associations and sequences as composite inputs to decision trees to determine relationships	
Clustering				Might be useful in creating composite clusters for inclusion in decision trees	
Regression				Use regression techniques to create linear composites for inclusion as inputs—a data reduction technique	

(continued)

Data Mining Methods Cross Impact Matrix					
	Association	Clustering	Regression	Decision Trees	Neural Networks
Decision Trees			Define strata for regression treatment Compute dummy variables Qualify variables in the equation (e.g., identify interactions) Impute missing values based on inputs with various levels of measurement		Prequalify variables for inclusion, including bins for categories Turn decision tree on predicted scores (and residuals) to assist in interpretation Turn decision tree on score residuals
Neural Networks				Fit and fine-tune unclassified observations	

Decision Trees in Analytical Model Development

The following example shows how a decision tree is used in a business-to-business marketing application. The analysis deals with network equipment sales. The display in Figure 6.12 shows current sales in the target geography. This display shows penetration rates based on current sales in a target U.S. sales region. Vendor sales data has been enriched by linking with Dun & Bradstreet data to show how sales are distributed according to enterprise size and vertical market. Penetration rates in large enterprises (over 10,000 employees) are good—74%. However, penetration rates in smaller enterprises (e.g., 500–1,000 employees) are lower —only 32%. Vertical market figures show that the best penetration is in large universities (26%); government, financial services, and health penetration rates taper off to less than 10%.

There are some important lessons here:

- Further penetration is likely only with lower-cost, lower-margin offers.
- Penetration into smaller enterprises in new vertical markets depends on indirect sales methods and leveraging partner relationships in these accounts.

The power of analytics comes from using known sales data to build a model that is applied against the universe of enterprises. Sales data comes from the vendor's sales data store. Universe data comes from third-party data vendors (in this case, Dun & Bradstreet; however, Harte-Hanks and InfoUSA also provide business-to-business data).

With known sales data (and the attributes of these sales), you can match the sales and their attributes with the attributes of the universe data to determine where the opportunities are (and how strong and what kind they are). To do this, you need a predictive model.

You should begin by using decision trees to look through the data. From the Dun & Bradstreet data, you can get information on potential key predictors of sales. This is basically corporate demographics, sometimes called firmographics. Decision trees are one of many prediction methods available in SAS Enterprise Miner. They are very useful in the early stages of forming predictive models because they provide a robust view into the relationships in the data.

Figure 6.10 shows the relationship between the enterprise size and the probability to purchase new technology. There is a strong positive relationship that shows that among small enterprises, new technology purchases are in the range of 30% of all enterprises. Among larger enterprises (e.g., those with more than 7,750 employees), 60% have new technology purchases. Marketing and sales planners can take advantage of this knowledge when planning sales campaigns. Predictive models can also take advantage of this knowledge and combine it with knowledge of other known relationships to form a multiple predictor model.

Figure 6.10: Illustration of Decision Trees Predicting New Technology (Business-to-Business)

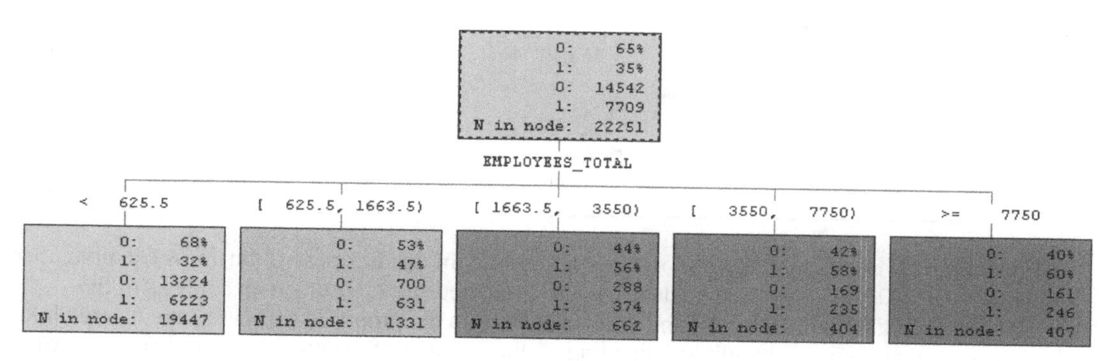

In this example, numerous decision trees were run to determine what strong relationships existed in the data. The following analytics-derived sales predictors were identified in the analysis:

- head count
- corporate location
- regional concentration
- business type
- PC estimator
- multi-site indicator (business knowledge)

After a good knowledge of the data is extracted using decision trees, it is useful to combine decision trees with other predictive approaches, specifically regression and neural networks. Combined predictive models, sometimes called ensembles, can produce better predictions. A combined approach is illustrated in Figure 6.11.

Figure 6.11: Illustration of the Combined Predictive Model to Compute Business-to-Business Propensity Scores

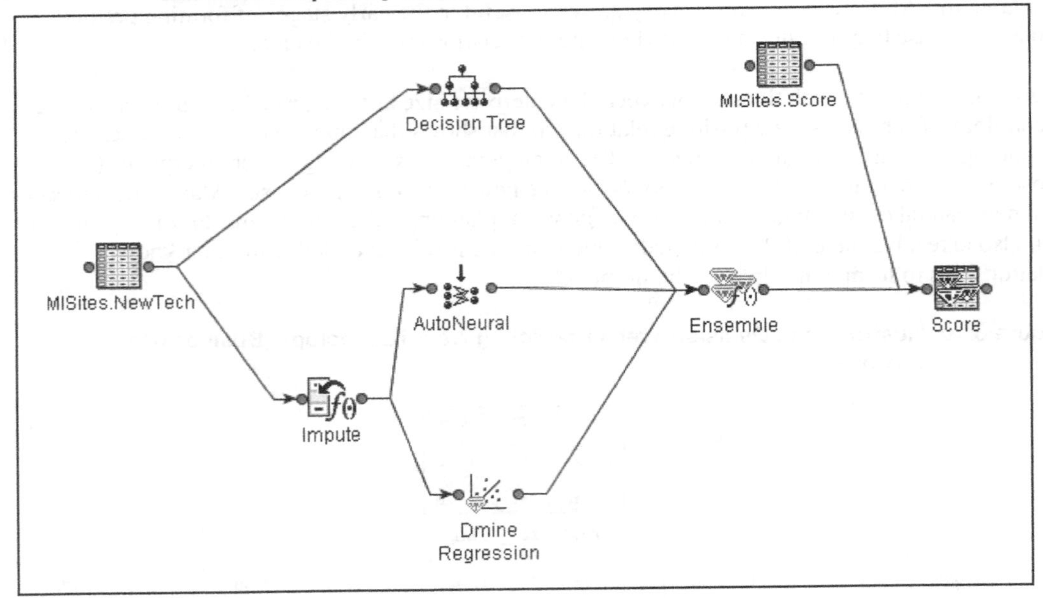

The final results of this model are shown in Figure 6.12. In this business-to-business example, there is a total of 15,309 customers in the data set. The predicted sales, based on an analysis of the current sales data as applied to all candidate businesses in the proposed sales area, are classed into high, medium, and low probability (depending on the strength of the combined predictive score). As shown in Figure 6.12, there were over 76,000 high-probability purchasers in this sales area. As also illustrated in Figure 6.12, general businesses (fewer than 250 employees) have the largest number of high-probability sales (38,348).

Figure 6.12: State Enterprise Footprint

State Enterprise Footprint (with site counts)
and Enterprise Penetration Rates

	Enterprises	Sites	Accounts	Penetration	Enterprise High Probability Purchase
Segments					
Enterprise 10k+	255	174462	**188**	74%	4332
Gen. Business (<250)	277828	410404	**6257**	2%	38348
Majors	10086	178168	**3058**	30%	33625
Majors < 500	5119	35727	**1063**	21%	
Majors 500 <1000	2571	32193	**824**	32%	
Majors 1000 < 2500	1521	39710	**680**	45%	
Majors 2500 < 5000	594	35165	**330**	56%	
Majors < 10000	281	35373	**161**	57%	
Sml Business (< 10)	1844573	1865348	**2511**	0%	38
Totals	2132742	2628382	**12014**		76343
Verticals					
Education	16895	38774	**1138**	7%	4777
Ed Services	8030	10360	**114**	1%	
Elem and Sec Sc	6278	22677	**642**	10%	
Jr. Colleges, D	877	1400	**150**	17%	
Univ. + 4yr col	830	2995	**215**	26%	
Vocatn	880	1342	**17**	2%	
Financial Services	76237	126989	**911**	1%	7002
Banking	9268	31808	**267**	3%	
Credit Unions	1640	2916	**168**	10%	
Insurance	40048	53593	**278**	1%	
Other, CC, Mortgage	20994	30044	**151**	1%	
Securities	4287	8628	**47**	1%	
Government	13285	46716	**664**	5%	4468
Local Municipal	1048	1697	**28**	3%	
National, Federal	468	8659	**21**	4%	
Public sector	11546	36082	**611**	5%	
Super national	223	278	**4**	2%	
Health	120255	142684	**999**	1%	4708
Horizontal	1906070	2273219	**8302**	0%	55388
Totals	2132742	2628382	**12014**		76343

These results show how decision trees can be used to explore the data before constructing predictive models, which might combine multiple predictive approaches. They also show the value of predictive modeling in general; for example, in this sales area, there were only about 15,000 customers, yet the predictive model indicated that there were as many as 76,000 high probability purchasers. These results can be used to construct sales campaigns to contact these highly probable customers.

The Use of Decision Trees in Rule Induction

Decision trees are often described as *inductive rule learners* because the structure of the decision tree begins inductively with the identification of one rule that partitions the main data set. The main data set is presented as the root node of the tree and typically contains a summary of the target field to be analyzed. The global structure of the tree is developed inductively—rule by rule—and is based on the operation of an iterative, one-step-at-a-time process.

The naming convention that has been adopted for the Rule Induction node is based on recognition of the inductive characteristic of decision trees. The specific algorithm that underlies the Rule Induction node is particularly well-adapted to dealing with the presence of relatively rare frequencies for certain values in the target field.

This strategy of successively defining rules on subsets and then removing those subsets is sometimes called *separate-and-conquer rule induction*, a phrase introduced by Giulia Pagallo and David Haussler (1990). The earliest examples that use this approach are AQ (Michalski, Mozetic, Hong, and Lavrac, 1986) and CN2 (Clark and Niblett, 1987). Those early algorithms were innovative but slow to execute. William W. Cohen at the School of Computer Science at Carnegie Mellon University developed a more practical algorithm, called ***Ripper,*** for large noisy data sets. ***Ripper*** is an example of a *sequential covering* approach in data mining. In this approach, we first find one rule that predicts a given outcome. When the rule is found, the data that is covered by this rule is removed from the training data; i.e., it is ripped out. As outlined in the paper that describes the algorithm, entitled "Fast Effective Rule Induction" (Cohen, 1995), the goal of this approach is to develop a rule learning algorithm that will perform efficiently on large noisy data sets. The criteria for identifying the rule is that it has high accuracy; i.e., as much as possible it should predict 100% of the observations that are matched by the rule. This process is unlikely to be able to cover a lot of observations. Once the observations that conform to the first rule are identified and then ripped out, the algorithm proceeds to find the next best rule, once again, as much as possible finding a high accuracy (but potentially low coverage) set of observations. These approaches, which might work well with the training data but are less likely to generalize well with new data sets, are termed *greedy*. So, greedy algorithms might satisfy the immediate urge to characterize training data well, but they do not tend to perform well in the long run.

We can see that this approach to rule induction is different than the normal method of working with decision trees. A typical rule in a decision tree uses a single variable to assign an observation to a branch. Here, at any given level of the tree, we find a node, a partitioning field, and—on the next level—the associated descendent leaves that result from the operation of the partitions, or splits. The decision tree is an unfolding, level by level, set of rules, where each rule corresponds to the values of one input variable. Specific outcomes and the corresponding rule for each variable are determined by the collapsing of input values together to form categories that predict the target.

In the case of ***Ripper***, for example, we see that a typical rule induction can use many variables to make a prediction about a specific subset of observations. This subset is designed to predict a pure target node, which contains only one value. *Ripper* is a fast way to create a rule induction to grow a tree and find a pure leaf. The branches that combine to produce a pure leaf form the terms of the combined rule. This combined rule covers the subset, so it determines the observations that are assigned to the leaf.

Iterative Removal of Observations

Once the subset that is covered by the rule is identified, then all the observations that are associated with this subset—as identified by the terms of the combined rule—are removed from the training data before the next iteration begins. In other words, once the rule is identified, then all observations that are associated with this rule are removed from the training data, even if they do not contain the predicted target values. This latter situation can occur when an inductive rule is selected in situations where the target is not 100% pure. The effect of this step is to remove potentially misleading training observations from the data.

Numerous versions of the algorithm have been developed. A version of rule induction has been developed at SAS that implements a variation of the ***Ripper*** algorithm. The SAS version of rule induction that is available on the Model tab employs a number of refinements that enable the Rule Induction node to provide robust, industrial-quality results across a variety of data sets. These recent refinements from SAS introduce the ability to create inductive rules by using neural networks and regression, in addition to decision trees. In summary, some of the refinements introduced by SAS include:

1. The ability to run an initial decision tree model against the data in an effort to find a rule that leads to increased purity in the first-level terminal nodes of the tree. This optional first step can be run without forcing the removal of the training cases that are covered by the rule.
2. The Rule Induction node provides for a "purity level" threshold value. In principle, rule induction looks for rules that discover complete separation between positive and negative cases in a node (i.e. the rule cleanly separates all positive and negative examples in the training set). Good rules that have less than a complete separation might exist, however, and the inclusion of a purity threshold that is less than 100% allows the algorithm to discover these rules as well.
3. Once all valid rules have been discovered, there might still be residual observations that could conceivably be classified using a standard classification method, such as a decision tree. In the Rule Induction node this is called the cleanup model.

   In the SAS version, both the initial model and the cleanup model can be run with alternative modeling approaches such as regression and neural networks.

The process is illustrated in Figure 6.13.

Figure 6.13: The Rule Induction Process

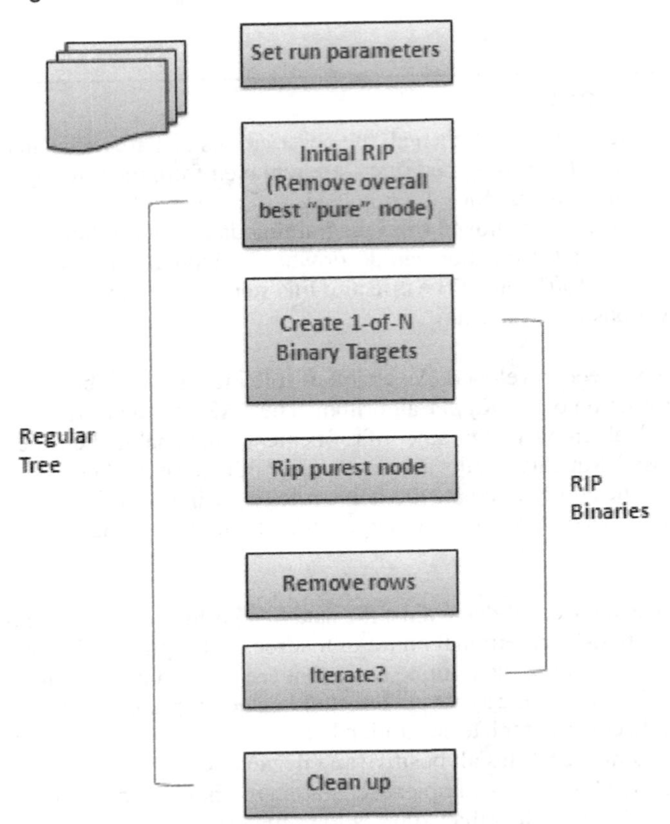

1. As shown in the top of the illustration, rule induction begins with all the training observations.
2. The rule induction process creates a binary model subset for each level of a target variable (shown in the figure as **Create 1-of-N Binary Targets**). This results in an exhaustive definition of the potential subsets that can be covered by the model.
3. Initializing the run parameters begins by setting the purity threshold for the binary rule induction part of the algorithm. This purity threshold can range from 80%–100%. In the run parameters, you can forego the initial Rip and instead run a normal decision tree, neural network, or regression process. In any case, the first procedure that is run selects the subset that contains the highest percentage of response categories. If the initial Rip has been selected, then rows that correspond to this node are removed from the training data. If the Rip is not selected, then all observations are left in the data.
4. The run parameters in the node also provide an option to operate on the rarest category first or the most frequent category first (rarest by default).
5. A tree model is run to find the largest node that meets the purity threshold that you set.

6. Rows in the training set for which the rare event is correctly classified are removed from the training data set. This forms a new training set with a reduced number of training observations.
7. Once again, binary targets are created for the remaining values in the training set.
8. Continue producing rules and eliminating training rows as long as purity criterion is satisfied.
9. Run a cleanup model of all of the unclassified rows. The cleanup model operates on all levels of the target.

Rule Induction Example

This example uses the Fisher Iris data provided in the SAS Enterprise Miner sample data sets. The training data is taken from the classic study described by Fisher (1936).

The Wikipedia entry about this data set (http://en.wikipedia.org/wiki/Iris_flower_data_set) notes: "The data set consists of 50 samples from each of three species of Iris (Iris setosa, Iris virginica and Iris versicolor). Four features were measured from each sample: the length and the width of the sepals and petals, in centimeters. Based on the combination of these four features, Fisher developed a linear discriminant model to distinguish the species from each other."

We use the default settings, as shown in Figure 6.14.

Figure 6.14: Rule Induction Default Settings (Decision Tree Cleanup)

Train	
Variables	...
Initial Ripping	Yes
Purity Threshold	100
Max. Number of Rips	16
Binary Models	Tree
Binary Order	Descending
Cleanup Model	Tree

Figure 6.15 provides a solution in decision-tree form. In this case, the decision tree display and the actual rule induction implementation are almost identical. This example, in particular, illustrates the high degree of correspondence between decision trees and rule induction.

Figure 6.15: Illustration of Rule Induction Using a 100% Node Purity Criterion (Illustrated as a Decision Tree)

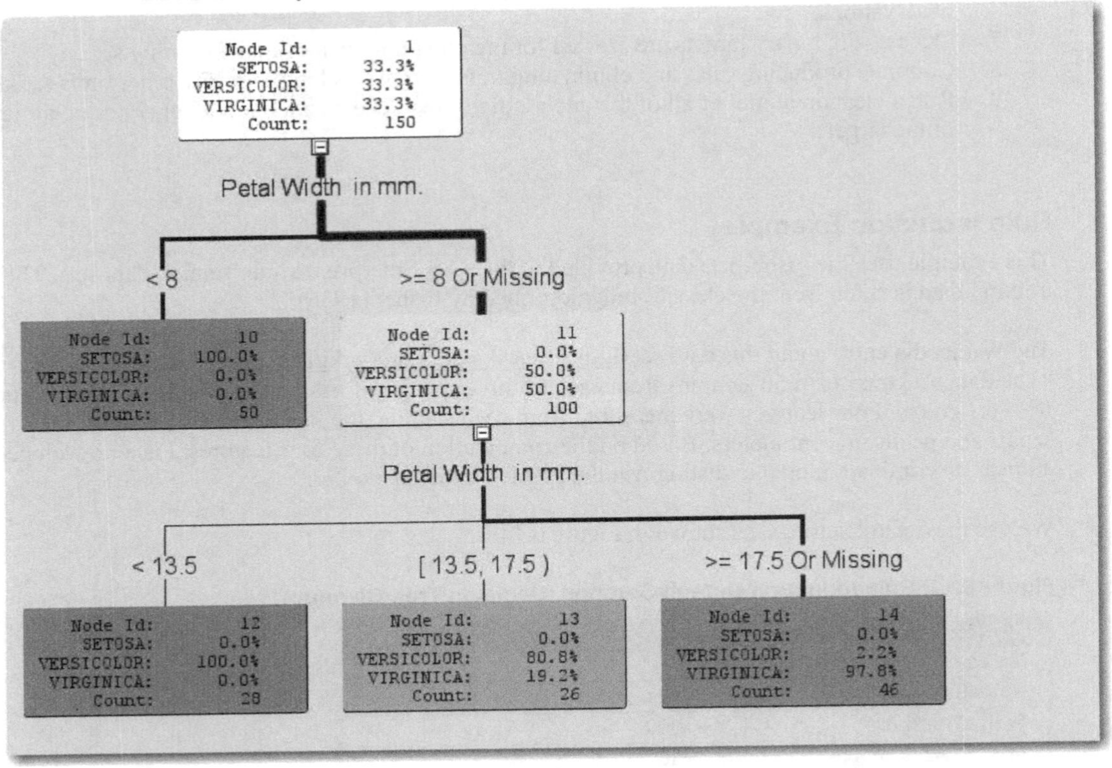

As shown in the figure, the data consists of three equal components consisting of setosa (50 records), versicolor (50 records), and virginica (50 records). When rule induction runs, the algorithm looks for a rule that can produce a pure node. The algorithm finds a rule, based on the Gini criterion, that segments the parent population into two segments based on the following criteria: Petal Width: less than 8 mm in one segment and 8 mm and greater (or missing) in the other segment. This rule results in the identification of one segment that is completely setosa. In the rule induction framework, this segment is ripped (or taken out of the sample), leaving the other segment that now consists of 50% – 50% versicolor and viginica segments.

We can see this reflected in the rule induction log (retrieve the log in the Results window by selecting View → Log).

Figure 6.16: Log Results (Shown for First RIP)

```
RIP1 Leaf Table: Threshold= 100
Leaf 2 was ripped from the model.

                    Predicted:     Predicted:     Predicted:
                     species=       species=       species=
Node       N          setosa       versicolor     virginica

  2       50         1.0000          0.0000         0.0000
  5       46         0.0000          0.0217         0.9783
  6       29         0.0000          1.0000         0.0000
  7       25         0.0000          0.8000         0.2000
```

In the next step, the remaining segment of the data is explored for a rule that can find a pure node. All records are combined together and a new search process is run to find a rule that can produce a pure node.

This produces the result shown in Figure 6.17.

Figure 6.17: Log Results (Second and Subsequent RIPs)

```
RIP2 Leaf Table: Threshold= 100
Leaf 4 was ripped from the model.

                    Predicted:     Predicted:
                     species=       species=
Node       N       versicolor     virginica

  3       46         0.0217         0.9783
  4       29         1.0000         0.0000
  5       25         0.8000         0.2000


RIP3 Leaf Table: Threshold= 100
No leaf was ripped from the model.

                    Predicted:     Predicted:
                     species=       species=
Node       N       versicolor     virginica

  3       46         0.0217         0.9783
  2       25         0.8000         0.2000
```

Here we see that Node 4 (identified as "Leaf 4" in the log) is the only pure node that is found, so this node is ripped from the data. This leaves 72 records that are mostly versicolor. Because the rule induction criterion is set to look for 100% pure nodes, the process stops here. If we refer to the decision tree display in Figure 6.15, we can see an example of the difference between rule induction and decision trees: In our current example for rule induction, we have identified three rules that result in the identification of three segments. In the decision tree, we see that there are two splits, which result in two rules that, by contrast, identify four segments. Notice that the right-hand branch results in a three-way branch.

Changing the Purity Threshold

The Rule Induction modeling node in SAS Enterprise Miner enables you to change the purity threshold for rule identification to values in the range 80%–100%. This relaxation of the purity threshold recognizes that there might be good rules that do not neatly cleave the sample into pure segments. To illustrate this approach with the Fisher data, we can use the settings displayed in Figure 6.18.

Figure 6.18: Adjusting the Purity Threshold

Train	
Variables	...
Initial Ripping	Yes
Purity Threshold	90
Max. Number of Rips	16
Binary Models	Tree
Binary Order	Descending
Cleanup Model	Tree

The only difference from the earlier example is that we are using a purity threshold of 90%. The decision tree diagram that results from this setting is shown in Figure 6.19.

Figure 6.19: Illustration of Rule Induction Using a 90% Node Purity Setting (Decision Tree Display)

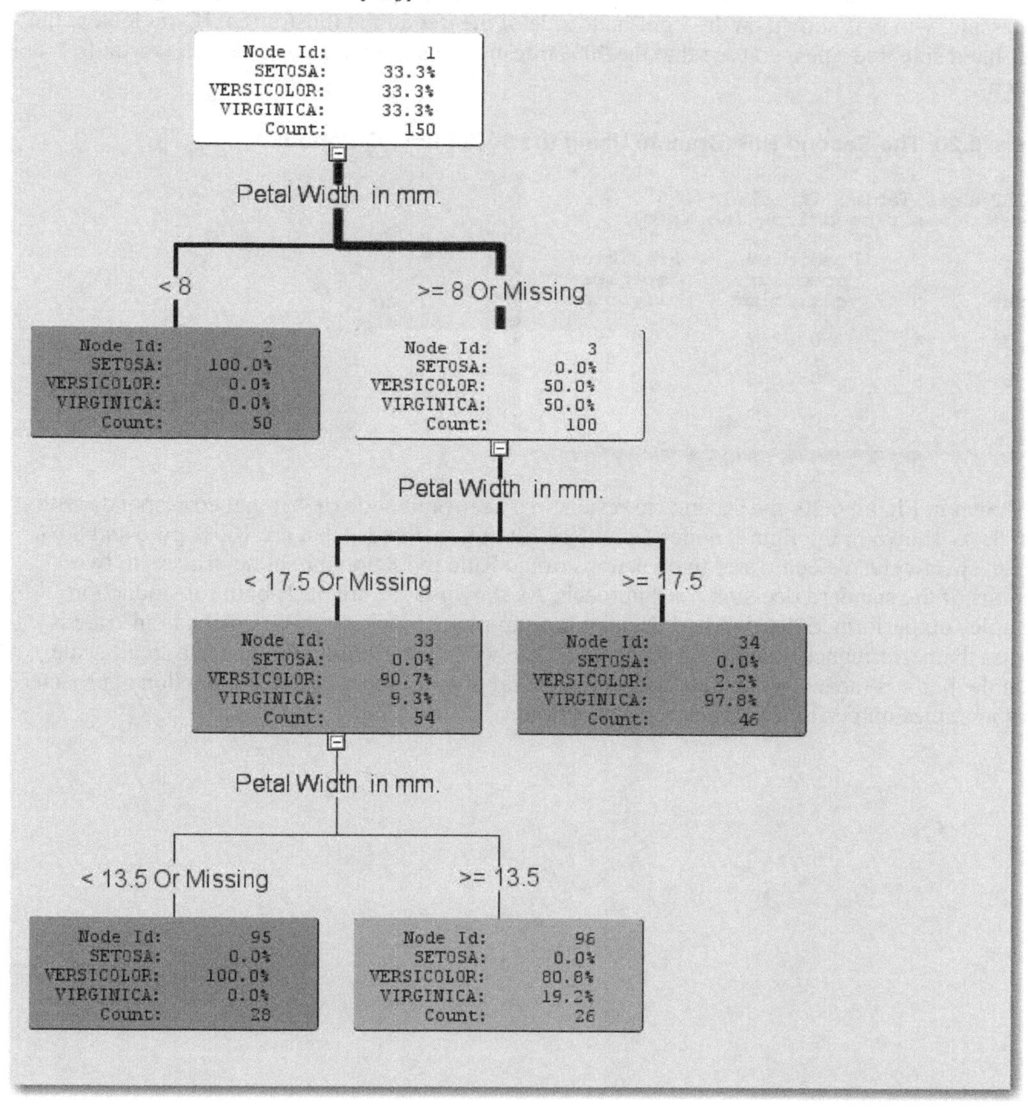

In Figure 6.19, we see that a pure node of 50 observations is first identified. These observations form a terminal node and so are effectively ripped out of the data in terms of subsequent processing, which is shown on the right-hand side of the tree in the illustration. If you look at the right-hand side tree rules, you see that the 90% threshold resulted in a RIP2 branch (second branch).

Figure 6.20: The Second RIP (Branch) Using the 90% Purity Criterion

RIP2 Leaf Table: Threshold= 90
Leaf 3 was ripped from the model.

Node	N	Predicted: species= versicolor	Predicted: species= virginica
3	46	0.0217	0.9783
4	29	1.0000	0.0000
5	25	0.8000	0.2000

As shown in Figure 6.20, the second rip results in a near-pure node (97%) that corresponds with Leaf 3. As shown in the figure, nodes 4 and 5 result in target nodes that are 100% pure and 80% pure, respectively.We compared two versions of the Rule Induction modeling process to two versions of the standard decision tree approach. As shown in Figure 6.21, both rule induction examples outperformed the standard decision tree approaches when "mis-classification" rate is used as the performance assessment measure (0.026 vs. 0.04). Admittedly, this is a small-scale example, but it is meant to illustrate the more general observation that a rule induction approach can sometimes outperform a standard decision tree.

Figure 6.21: Comparison of Rule Induction to "Classic" Decision Tree

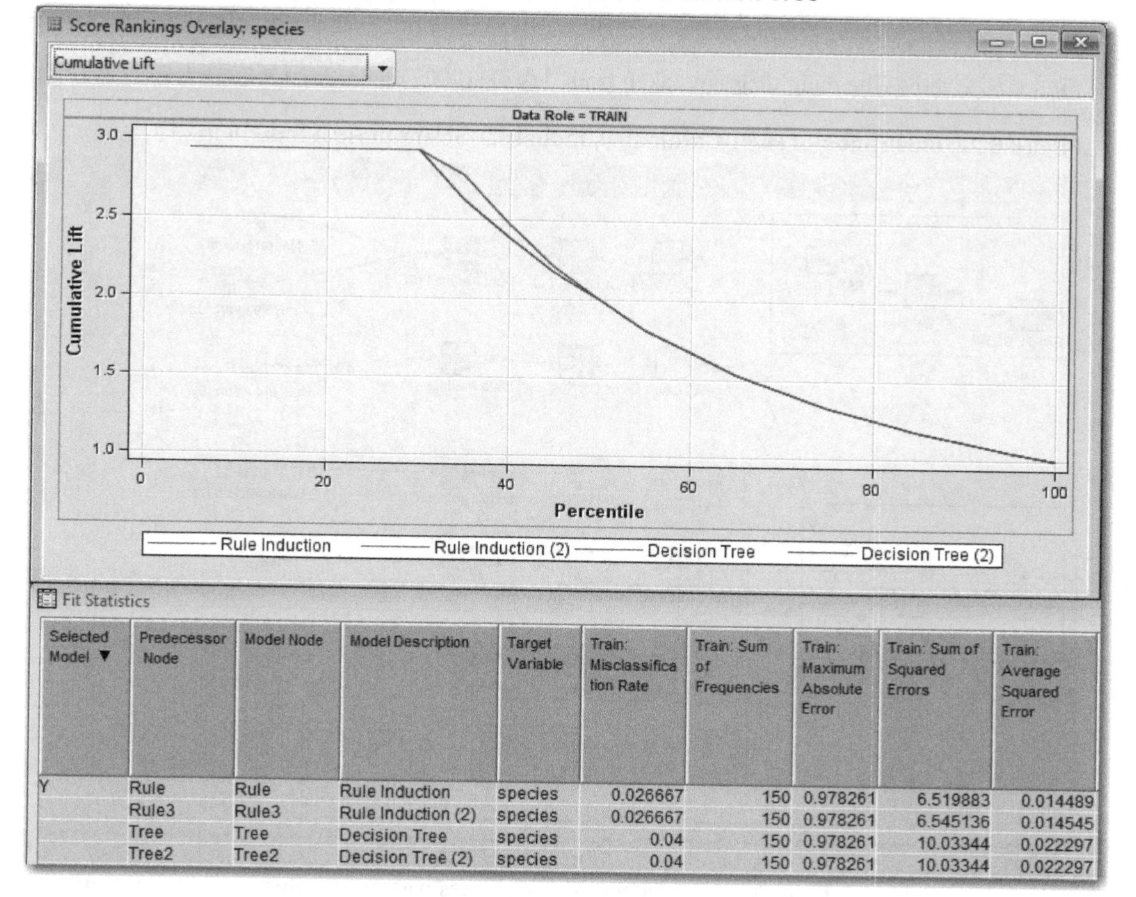

Selected Model ▼	Predecessor Node	Model Node	Model Description	Target Variable	Train: Misclassification Rate	Train: Sum of Frequencies	Train: Maximum Absolute Error	Train: Sum of Squared Errors	Train: Average Squared Error
Y	Rule	Rule	Rule Induction	species	0.026667	150	0.978261	6.519883	0.014489
	Rule3	Rule3	Rule Induction (2)	species	0.026667	150	0.978261	6.545136	0.014545
	Tree	Tree	Decision Tree	species	0.04	150	0.978261	10.03344	0.022297
	Tree2	Tree2	Decision Tree (2)	species	0.04	150	0.978261	10.03344	0.022297

A Robust, Industrial Quality Example

There are situations where rule induction makes a substantial difference when compared to the decision tree approach. Results reported by deVille (2007) illustrate the use of rule induction in the solution of a difficult problem of using text fields to classify warranty repairs. The main difficulty was the large number of warranty actions—over 1900. Ideally, this requires a classification algorithm that can work with a target field with over 1900 unique categories … an unusual, but increasingly more frequent requirement.

The problem was compounded by the low frequency of target categories in any one of the fields (the in-category frequencies ranged from 0%–to 2% in the 20 most common warranty categories, to generally 1% and lower in the majority of categories). The full approach is described in greater detail in the source cited above and in the associated US Patent (2009). For purposes of this example, only the operation of the rule induction part of the successful algorithm is described here.

Figure 6.22 shows the main diagram, taken from deVille (2007).

Figure 6.22: Illustration of High-Cardinality, Industrial-Strength Rule Induction

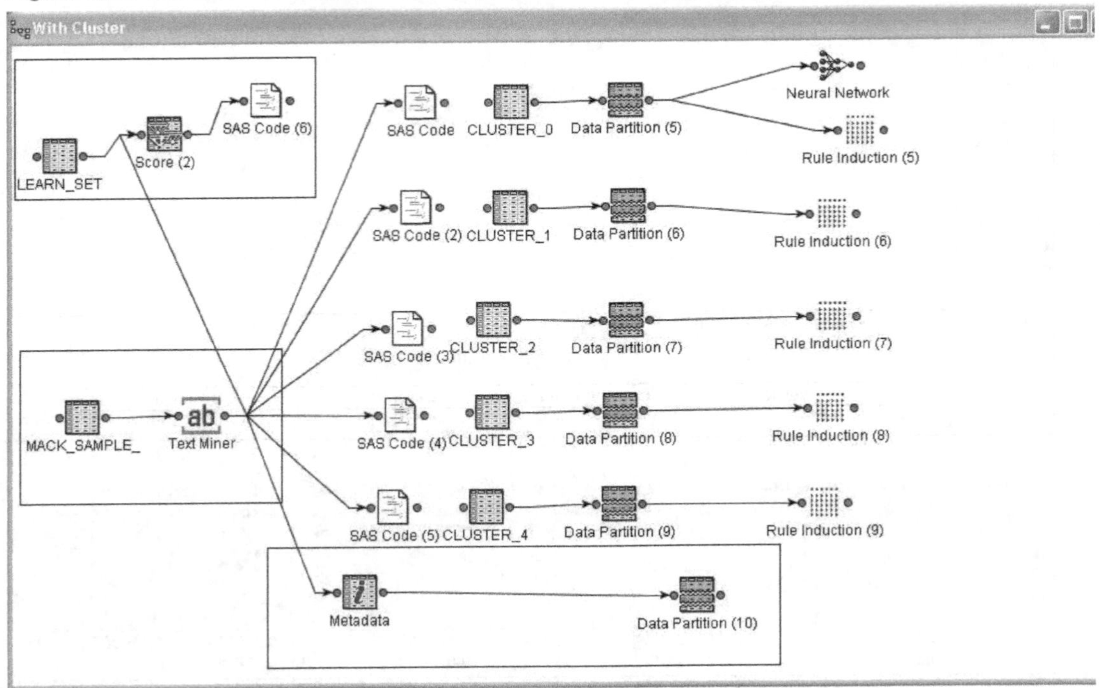

A key strategy in classifying so many categorical outcomes was to first cluster similar outcomes together. This is shown in the figure by the appearance of the five Cluster nodes. This step means that it is not necessary to classify all the over 1900 categories at once. We can classify categories in segments, which are shown here as the clusters.

After a group of targets were clustered, the next step was to induce a rule for each of the constituent targets so as to generate one or more rules to classify each of the target categories in the study.

Table 6.2: Example Target Field Values in Rule Induction Problem

Target	Train Count	Train Percent	Valid Count	Valid Percent
2389	1132	3.7	5486	3.68
2361	911	2.98	392	2.97
2154	788	2.57	338	2.56
2231	777	2.54	335	2.54
1222	742	2.44	321	2.43
2341	746	2.44	321	2.43
2380	721	2.36	310	2.35
2100	667	2.18	287	2.18
2510	652	2.13	279	2.12
2132	628	2.05	271	2.05
2360	623	2.04	269	2.04
2112	616	2.01	264	2.00
2131	609	1.99	263	1.99
1721	580	1.89	250	1.90
2371	558	1.82	240	1.82
2155	554	1.81	237	1.80
2121	516	1.69	222	1.68
2372	513	1.68	220	1.67
1524	479	1.57	208	1.58

.
.
.

(Partial Results Table Continues)

Full details of the operation of the algorithm are discussed in the above-mentioned sources. As shown in Figure 6.23, both quantitative fields (such as miles driven) and qualitative fields (such as claim type) were used in the prediction. Singular Value Decomposition terms, which are highly summarized numerical representations of the textual contents in the warranty description field, are also used in the predictive model. The combination of particular numeric fields and singular value vectorization expressions turns out to be a very good predictor of unique warranty actions—particularly when these rule combinations are identified using the Rule Induction node.

Figure 6.23: Field Attributes in Large-Scale Rule Induction

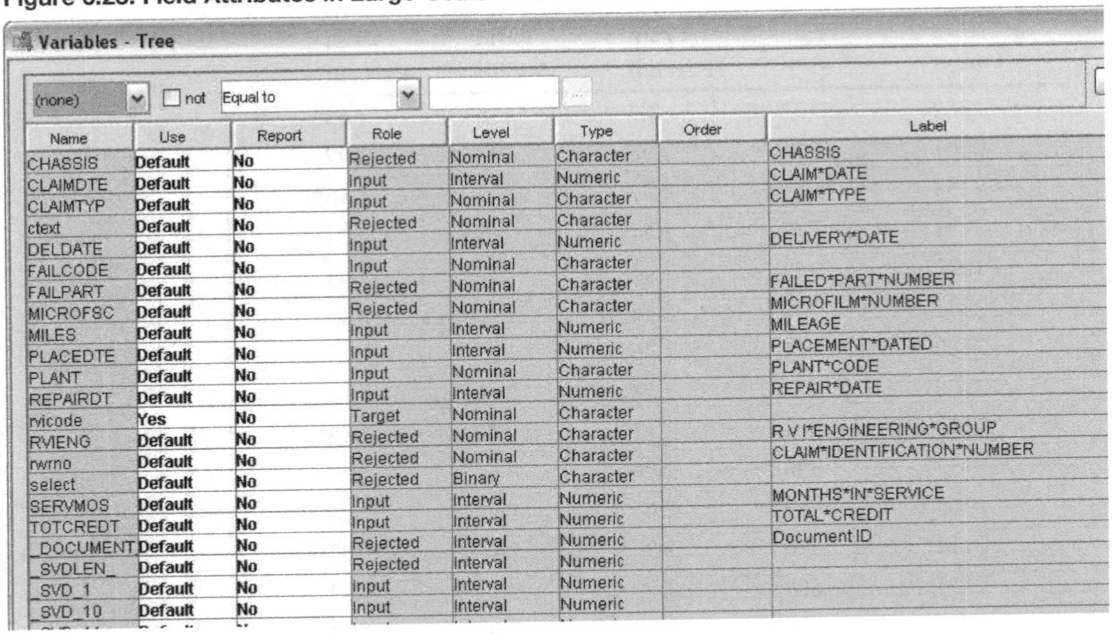

Variables - Tree

(none) | not | Equal to

Name	Use	Report	Role	Level	Type	Order	Label
CHASSIS	Default	No	Rejected	Nominal	Character		CHASSIS
CLAIMDTE	Default	No	Input	Interval	Numeric		CLAIM*DATE
CLAIMTYP	Default	No	Input	Nominal	Character		CLAIM*TYPE
ctext	Default	No	Rejected	Nominal	Character		
DELDATE	Default	No	Input	Interval	Numeric		DELIVERY*DATE
FAILCODE	Default	No	Input	Nominal	Character		
FAILPART	Default	No	Rejected	Nominal	Character		FAILED*PART*NUMBER
MICROFSC	Default	No	Rejected	Nominal	Character		MICROFILM*NUMBER
MILES	Default	No	Input	Interval	Numeric		MILEAGE
PLACEDTE	Default	No	Input	Interval	Numeric		PLACEMENT*DATED
PLANT	Default	No	Input	Nominal	Character		PLANT*CODE
REPAIRDT	Default	No	Input	Interval	Numeric		REPAIR*DATE
rvicode	Yes	No	Target	Nominal	Character		
RVIENG	Default	No	Rejected	Nominal	Character		R V I*ENGINEERING*GROUP
rwrno	Default	No	Rejected	Nominal	Character		CLAIM*IDENTIFICATION*NUMBER
select	Default	No	Rejected	Binary	Character		
SERVMOS	Default	No	Input	Interval	Numeric		MONTHS*IN*SERVICE
TOTCREDT	Default	No	Input	Interval	Numeric		TOTAL*CREDIT
DOCUMENT	Default	No	Rejected	Interval	Numeric		Document ID
SVDLEN	Default	No	Rejected	Interval	Numeric		
_SVD_1	Default	No	Input	Interval	Numeric		
_SVD_10	Default	No	Input	Interval	Numeric		

In this approach, there might be a rule identified for each unique category in the target because the rule induction process will continue as long as a rule is found that isolates 80% of the observations in a given category to a descendent leaf node of the rule before running the cleanup model.

As reported in the findings, this approach usually resulted in an accuracy rate of about 60%. Often the results were closer to 80% accuracy. In all cases, rule induction performed better than an associated decision tree. This is one of many examples that demonstrate the utility of rule induction for rare event prediction and classification.

Conclusion

Two of the many themes explored in this book relate to the synergy and complementarities found between decision trees and business intelligence, and the synergy and complementarities between decision trees and data mining tools and predictive modeling techniques. It seems likely that the complementarities will continue in these respective areas and, in so doing, will likely lead to increased synergy and integration in the future. Further developments in these areas include those discussed in the following sections.

Business Intelligence

Decision tree drill down through any face of any cube constructed through the multidimensional data interface.

Drill down provides the ability to retrieve the underlying detail data that the cube surface summaries are based on. Drill down enables the analyst to examine data in any of the multidimensional segments formed by the cube. The analyst can reveal predictive and classification structure on this detail data through decision tree execution against the detail data.

Dimensional aggregation in-line with decision tree methods.

It should be possible to express any dimensional view in an aggregated way using decision tree algorithms to collapse across dimensions with one or more of the measures being displayed. This provides a level of data summarization and data effect identification in what is otherwise a manual operation done “by eyeball.”

Decision trees in cube form for subsequent display, viewing, and analysis.

Just as data cubes can be pre-stored or pre-computed structures, so should decision trees. This ability is available in some applications. For example, decision tree results are used to determine the importance of various faces of the cubes. Decision tree results and input value clusters that are determined by decision tree methods can be used to produce collapsed values for dimensional cube faces.

Data Mining

The use of multiple decision tree approaches.

Multiple decision tree approaches that use a variety of boosting and bagging techniques are emerging. These approaches will be more routine in the future, especially as computational power increases to support this kind of decision tree growth in real time. Decision trees grown this way retain ease of use and the presentation characteristics that they share with business intelligence methods, while more closely resembling the other classic data mining methods, such as regression, neural networks, and cluster analysis.

More information about inputs.

With the growth of metadata repositories and metadata definitions, data sets are acquiring more information that defines the characteristics of inputs. For example, in addition to characteristics such as data type (integer, character, or numeric), you might expect to see the data element origin (such as customer table or transaction table). This could make it possible to associate a theoretical role and method to fields of data. Many applications use custom-defined fields to guide the construction of business intelligence dimensional reports. Likewise, extended metadata can be used to introduce business rules in the construction of decision trees. This has a number of benefits (for example, decreased computation and automated construction of reports and decision structures that are relevant to specific business uses).

Glossary

adjusted significance
: a significance measure that has been adjusted for the number of tests that were carried out in order to determine the level of significance. This adjustment prevents the identification of statistically significant results by chance.

algorithm
: a sequence of actions that performs a task. A procedure for solving a recurrent mathematical problem.

analytical model
: a structure and process for analyzing a data set. For example, a decision tree is a model for the classification of a data set.

anomalous data
: data that result from errors (for example, data entry errors) or that represent unusual events. Anomalous data should be examined carefully because they might carry important information.

ANOVA
: ANalysis Of Variance. A procedure used to detect statistically significant effects induced by an independent variable on a continuous dependent variable. The ANOVA procedure employs an F-test to measure the differences between a given set of population means, where F = Mean Square for Treatments divided by the Mean Square for Error (F = MST/MSE). See F-ratio statistic.

artificial neural network
: non-linear predictive models that learn through training and resemble biological neural networks in structure.

ASCII
: American Standard Code for Information Interchange. A standard set of character codes established by the American National Standards Institute that enables you to transmit text between computers or between a computer and a peripheral device.

assessment plot
: a line graph that shows a plot of the accuracy of the decision tree for various subtrees.

bagging
: a method that resamples the training data to create a pooled estimate of the target. Various decision trees are grown independently, and a group vote is used to produce the target estimate. Bagging stands for bootstrap aggregation and was developed by Leo Breiman.

binary variable
: a variable that takes only two distinct values. A binary variable is the most basic form of measurement indicating the presence or absence of some characteristic.

Bonferonni adjustment
: a conservative adjustment that is applied to a test of significance in order to compensate for the number of statistical or mathematical operations that are performed in advance of a specific statistical test. These adjustments are designed to ensure that the test statistic conforms to the assumptions necessary for its calculation (for example, no dependencies between one test and another).

boosting
: the process of resampling the data to form a succession of decision trees many times to form one average estimate for the target. Each time, the data is used to grow a tree and the accuracy of the tree is computed. The successive samples are adjusted to accommodate previously computed inaccuracies. Because each successive sample is weighted according to the classification accuracy of previous models, this approach is sometimes called stochastic gradient boosting.

case
: a collection of measurements regarding one of numerous entities represented in a data set. Synonyms: observation, record, example, pattern, sample, instance, row, vector, pair, tuple, fact.

case weight
: a positive numeric variable that serves as a multiplier to magnify the contribution of each line of data to an analysis. There are three kinds of case weights: frequencies, sampling weights, and variance weights.

categorical field/variable
: a variable that can assume only a limited number of discrete values. Shoe size and hair color are examples of categorical variables. A variable that lies in a nominal measurement space is sometimes called a qualitative, discrete, non-metric, or classification variable.

category
: one of the possible values of a categorical variable. Synonyms: class, label.

CHAID
: Chi-square Automatic Interaction Detection. A method of segmenting a file. The method is applied to a discrete response variable.

chi-squared test

a test measuring the statistical association between two categorical variables.

class variable

in such fields as data mining, pattern recognition, and knowledge discovery, a class variable is a categorical target variable. Classification is the process of assigning cases to categories of a target variable. In traditional research methodology, class variables are categorical variables and can be used as either an input or a target.

classification

the process of dividing a data set into mutually exclusive groups such that the members of each group are as close as possible to one another, and different groups are as far as possible from one another. Distance is measured with respect to the specific variables you are trying to classify. For example, a typical classification problem is to divide a database of companies into groups that are as homogeneous as possible with respect to creditworthiness.

classification model

a model that predicts the class value of a categorical—or class—target. See class variable and predictive model.

clustering

the process of dividing a data set into mutually exclusive groups such that the members of each group are as close as possible to one another, and different groups are as far as possible from one another. Distance is measured with respect to all available variables.

column

contains a field of information in which each new column entry corresponds to a new row of data. In database terms, there can be many columns of data, each containing many rows of values. Row and column data attributes are familiar as database terminology and are sometimes referred to as cases and variables in research data settings.

contingency table analysis

tabular analysis, which is the analysis of crosstabulations.

continuous field

a field that has a numeric or ordered range of values such as temperature readings (e.g., 25, 26, 27, ...).

correlation

a statistical measure or the association (or co-relation) between two fields of data.

CRT

Classification and Regression Trees. A decision tree technique developed by Brieman, et al. (1984) and used for classification of a data set. It employs a grow-and-prune strategy to develop a right-sized tree and associated set of rules. Branches are formed by creating two-way splits.

database
: closely related information that has been gathered together. Most databases consist of fields, which contain units of information, and records, which contain sets or collections of fields. In general, fields are stored in columns and records are stored in rows.

data cleansing
: the process of ensuring that all values in a data set are consistent and correctly recorded.

data mining
: the extraction of hidden predictive information, typically from large databases that are often assembled from disparate sources.

data navigation
: the process of viewing different dimensions, slices, and levels of detail of a multidimensional database. See dimensional cube.

data visualization
: the visual interpretation of complex relationships in multidimensional data through scatter plots, dimensional cubes, and contour plots, for example.

data warehouse
: a system for storing and delivering numerous sources of data in a unified and accessible location.

decision tree
: a tree-shaped structure that represents a set of decisions. These decisions generate rules for the classification of a data set. See CRT and CHAID.

dependent variable
: the field that is analyzed as a function of other fields or variables in a data set. Also called the target field.

dimension
: in a flat or relational database, each field in a record represents a dimension. In a multidimensional database, a dimension is a set of similar entities; for example, a multidimensional sales database might include the dimensions product, time, and city.

dimensional cube
: an interactive analytical processing technique that originally referred to database applications that enable users to view, navigate, manipulate, and analyze databases as multidimensional entities. The approach has been incorporated into SQL to produce multidimensional summaries, and is now used for a variety of multidimensional reports and data manipulations that are based on dimensional cubes.

example
: a member of a training set, with measures for various attributes used to derive a decision tree structure. Equivalent to a subject, record, or observation.

exhaustive partitioning
: an alternative to standard CHAID branch grouping methods that is more likely to find the partitioning with the highest level of significance (because more groupings of values with respect to the dependent variable are formed). The partitions that are formed are empirically stronger than heuristically derived partitions. Decision trees formed using the exhaustive partitioning method tend to have more branches than those formed using the original method. Developed by Biggs, et al. (1991).

exploratory data analysis
: the use of graphical and descriptive statistical techniques to learn about the structure of a data set, usually as a preliminary step to predictive modeling.

extrapolation
: the scoring or generalization for values of observations outside or beyond a given training data set, typically on the basis of values or functions taken from other inputs in the training data set. Often used for predicting likely values for new observations.

field
: a column that you label in your database that contains the same kind of information for each record.

floating
: a branch-clustering option originally developed by Kass that allows the missing values of an ordered field to group with other values in the field that they most closely resemble (i.e., they have a similar effect on the dependent variable as the ones they are grouped with).

F-ratio statistic
: a value calculated as part of the ANOVA procedure. The larger this number is, the greater the distance between the means or average values of the nodes in the split. See ANOVA.

generalization
: the ability of a model to compute good outputs from input data not used during training. Synonyms: interpolation and extrapolation, prediction.

genetic algorithm
: an optimization technique that uses processes such as genetic combination, mutation, and natural selection in a design, based on the concepts of natural evolution.

heuristic partitioning
: a method of partitioning field data that provides optimal branching based on heuristics or statistical rules of thumb. It is less time consuming than exhaustive partitioning and tends to produce fewer branches and more compact trees than the exhaustive approach.

ID3
: a machine-learning algorithm.

independent variable
: one of potentially many fields or variables that are used to describe, predict, or explain variability in a dependent or target field. Independent variables are usually called inputs in a data mining context because the input value influences the outcome of the model describing the relationship between the input and target.

induction
: a method of proving statements about an ordered data set. Induction reasons from particulars to generals or from the individual to the universal. Synonym: inference.

input
: a variable used to predict or guess the value of the target variables. Synonyms: independent variable, predictor, regressor, explanatory variable, carrier, factor, covariate.

interaction effect
: an effect on the relationships between two (or more) variables where the direction of the relationship (i.e. positive or negative) depends on the value of another variable. An example of interaction effect is the relationship between weight and blood pressure changes for different age groups.

interpolation
: the scoring or generalization for values of observations in a given training data set, typically on the basis of values or functions taken from other inputs in the training data set. Often used for estimating missing values.

interval
: a defined range of values.

interval boundary
: a breaking point in a continuous field that divides the field into intervals.

interval variable
: a numeric variable for which arithmetic operations with values are informative. An interval level of measurement means that the observed levels are ordered and numeric and that any interval of one unit on the scale of measurement represents the same amount, regardless of its location on the scale. Typical interval scales include income and temperature.

leaf
: the bottom or final nodes in a decision tree.

linear model
: an analytical model that assumes linear relationships in the coefficients of the variables being studied.

linear regression

a statistical technique used to find the best-fitting linear relationship between a target (dependent) variable and its predictors (independent variables).

–log (p)

see logworth.

logistic regression

a linear regression that predicts the proportions of a binary category target variable, such as type of customer, or has attribute versus does not have attribute.

logworth

a transformation of the normal method of presenting significance that takes a negative log of the significance in order to express greater levels of significance in larger numbers (so that the magnitude of the significance is reflected in the magnitude of the number).

measurement

the process of assigning numbers to objects such that the properties of the numbers reflect some attribute of the objects.

measurement level

one of several different ways in which properties of numbers can reflect attributes of objects. The most common measurement levels are nominal, ordinal, interval, log-interval, ratio, and absolute. For details, see the measurement theory FAQ at ftp://ftp.sas.com/pub/neural/measurement.html.

metric

supports arithmetic operations. See interval.

missing value

a value that is absent from a field. Missing values are represented as a period (.) in SAS.

model

a general term that describes a conceptual representation of some phenomenon typically consisting of symbolic terms, factors, or constructs that can be rendered in language, pictures, or mathematics. Models include formulas or algorithms for computing outputs from inputs. A statistical model also includes information about the conditional distribution of the targets given the inputs. See trained model.

multidimensional database

a database designed for online analytical processing (dimensional cube). The database is structured as a multidimensional hypercube with one axis per dimension.

multiprocessor computer

a computer that includes multiple processors connected by a network. See parallel processing.

nearest neighbor
: a technique that classifies each record in a data set based on a combination of the classes of the k records most similar to it in a historical data set (where k is greater than or equal to 1). Synonym: k-nearest neighbor.

node
: a location defined by branch attributes on a tree. The root node is the initial node displayed in a decision tree. All branches originate at the root node. The nodes on the bottom-most branches of the tree are terminal nodes or leaves.

noise
: an unpredictable variation, usually in a target variable. For example, if two cases have identical input values but different target values, the variation in those different target values is not predictable from any model that uses only those inputs; hence, that variation is noise. Noise is often assumed to be random. In that case, it is inherently unpredictable. Whereas noise prevents target values from being accurately predicted, the distribution of the noise can be estimated statistically, given enough data.

nominal variable
: a numeric or character categorical variable in which the categories are unordered and the category values convey no additional information beyond category membership.

nonlinear model
: an analytical model that does not assume linear relationships in the coefficients of the variables being studied.

null category
: a category that has no corresponding observation in a field displayed in a descendent node of the decision tree.

observation
: a data record or subject of a given collection of data in which one or more attribute measures are taken and recorded for each unit of analysis.

operational data
: a type of data to be scored in a practical application, containing inputs but not target values. Scoring operational data is the main purpose of training models in data mining. Synonym: scoring data.

operationalize
: to assign representations such as numeric tokens or concise term relations to conceptual entities.

ordered
: a clustering option to collapse input values that treats a set of values as an ordered sequence, and that allows only adjacent values to be grouped together.

ordinal
: a method of measurement whereby adjacent values are ordered. Typically, the ordering is monotonic such that each higher level adjacent category is at least as great as the lower category and might be greater by some measurement.

ordinal variable
: a numeric or character categorical variable in which the categories are ordered, but the category values convey no additional information beyond membership and order. In particular, the number of levels between two categories is not informative, and for numeric variables, the difference between category values is not informative. The results of an analysis that includes ordinal variables are typically unchanged if you replace all the values of an ordinal variable with different numeric or character values as long as the order is maintained, although some algorithms might use the numeric values for initialization.

outlier
: a data item whose value falls outside the bounds that enclose most of the other corresponding values in the sample. An outlier might indicate anomalous data. Outliers should be examined carefully; they can carry important information.

output
: a variable computed from the inputs as a prediction or guess of the value of the target variables Synonyms: predicted value, estimate, y-hat.

parallel processing
: the coordinated use of multiple processors to perform computational tasks. Parallel processing can occur on a multiprocessor computer or on a network of workstations or PCs.

parameter
: the true or optimal value of the weights or other quantities (such as standard deviations) in a model.

partitioning
: the act of breaking up a set of field values into discrete groups based on similarity with respect to a dependent variable as determined by a test of statistical significance.

pattern
: a set of relationships between fields of data typically derived through statistical methods, as in predictive modeling. Typically, the emphasis is on the display of the pattern as opposed to the prediction.

PMML
: Predictive Modeling Markup Language. PMML describes data mining models in Extensible Markup Language (XML), a universal format that describes structured documents and data designed by W3C group (http://www.w3c.org). The format was designed by the Data Mining Group (http://www.dmg.org) and enables researchers and commercial users to carry out various data mining tasks in a universal notation that is shared across environments. Typically,

these environments employ proprietary standards that would otherwise make interoperation difficult, if not impossible.

population
: the set of all cases that you want to be able to generalize to. The data to be analyzed in data mining are usually a subset of the population.

posterior probability
: the probability of a target category (taken from the distribution of categories in the target field) that is calculated after selecting the target field based on some prior value (typically a branch value in a decision tree).

predictive model
: a model with a target or outcome field or variable that is shown to be a function of one or more input or predictor fields or variables. Outcomes can be categorical (buy/no buy) or continuous (dollars spent; time spent). With categorical outcomes, the models are called classification models. With continuous outcomes, they are called regression models.

prospective data analysis
: a process that predicts future trends, behaviors, or events, based on historical data.

qualitative
: a process or entity that is defined in qualitative or non-exacting forms of measurement.

quantitative
: a process or entity that is defined in quantitative, numerically based terms.

random forest
: a collection of multiple decision trees that produce an average estimate for the target. In each node, a branch search is made on a random set of inputs instead of on the full set of inputs. Each decision tree in the random forest is grown on a bootstrap sample of the training data set.

ratio variable
: a numeric variable for which ratios of values are informative. In SAS Enterprise Miner, ratio and higher-level variables are not generally distinguished from interval variables because the analytical methods are the same. However, ratio measurements are required for some computations in model assessment, such as profit and ROI measures.

record
: a piece of information contained in a database that comprises an entry for each field in the database. For example, an employee database contains a record for each employee.

retrospective data analysis
: a process that provides insights into trends, behaviors, or events that have already occurred.

root node
: the node at the very top of a hierarchical decision tree display. In this node, values in the dependent variable are also represented.

row
: the second dimension—along with column—of a traditional table. Because data sets are usually stored in tables, the observations, or examples, that are captured by the data set are considered to be rows.

rule induction
: the extraction of useful if-then rules from data that is based on statistical significance.

sample
: a subset of the population that is available for analysis.

scoring
: a method of applying a trained model to data to compute outputs. Synonyms: running (for neural nets), simulating (for neural nets), filtering (for decision trees), interpolating or extrapolating.

signal
: a predictable variation in a target variable. It is often assumed that target values are the sum of signal and noise, where the signal is a function of the input variables. Synonyms: function, systematic component.

significance
: a measure of the strength of a relationship between sample elements, based on statistical probability.

split
: a partition in a set of field values.

standard deviation
: the square root of the variance. It is the measure of the level of variability in a collection of data. The larger the number, the greater the variability.

stochastic gradient boosting
: see boosting.

subtree
: a subset of the full decision tree, created by pruning one or more branches up from the bottom of the tree. Subtrees always contain the root node.

supervised learning
: an environment in which the goal is to predict or classify the value of an outcome or target measure based on a number of input measures.

surrogate
: predictive information that is held in a field, which is closely associated with the field that is being used as an input to form a branch. If a value is missing for an input, then a surrogate can be used to estimate the likely value of the input at that point in the decision tree.

target
: a field or variable that is being examined, estimated, or described by the data mining model or process. It is synonymous with a dependent variable in a statistical analysis or a modeled outcome. The target variable value is known in some currently available data, but will be unknown in some future/fresh/operational data set. You want to be able to predict or guess the values of the target variables from other known variables. Synonyms: dependent variable, response, observed values, training values, desired output, correct output, outcome.

test data
: data that contains input and target values. Test data is not generally used during model training, but is instead used to estimate generalization error. Test data is designed to provide an estimate of model performance in novel situations, and is ideally an independent collection of data that is separate from the data used in training the model.

time series analysis
: the analysis of a sequence of measurements made at specified time intervals. Time is usually the dominating dimension of the data.

trained model
: a specific formula or algorithm for computing outputs from inputs, with all weights or parameter estimates in the model chosen via a training algorithm from a class of such formulas or algorithms designated by the model. Synonym: fitted model.

training
: the process of computing good values for the weights in a model, or, for tree-based models, choosing good split variables and split values. Synonyms: estimation, fitting, learning, adaptation, induction, growing trees.

training data
: data that contains input and target values used for training to estimate weights or other parameters. This data is used to develop the data mining model. The notion of training derives from a machine-learning approach whereby the underlying development model mimics the extraction of knowledge from data through the use of lines of data as training instances.

twoing
: a node-partitioning technique that segments the classes in a node into two groups by combining classes together that form up to 50% of the data.

unsupervised learning
: an environment in which there is no outcome measure. The goal is to describe the associations and patterns among a set of input measures.

validation data
: data that contains input and target values used indirectly during training for branch selection and for determining when to form terminal nodes.

variable
: an item of information represented in numeric or character form for each case in a data set. Both targets and inputs are variables. Synonyms: column, feature, attribute, coordinate, measurement.

variance
: a measure of the range of values in a distribution that also combines a measure of the density. Variance is sometimes referred to as the second moment around the mean. Variance is the expected value of the square of the deviations of a random variable from its mean value.

weight
: a numeric value used in a model that is usually unknown or unspecified prior to the analysis. Weights can be estimated by the model or can be used in computing model results. Synonyms: estimated parameters, estimates, coefficients, betas.

References

Alexander, W. P., and S. D. Grimshaw. 1996. "Treed Regression." *Journal of Computational and Graphical Statistics* 5:156-175.

Amit, Y., and D. Geman. 1997. "Shape Quantization and Recognition with Randomized Trees." *Neural Computation* 9, no. 7:1545-1588.

Barlow, T., and P. Neville. 2001. "A Comparison of 2-D Visualizations of Hierarchies." *Proceedings of the IEEE Symposium on Information Visualization 2001 (INFOVIS'01)*. San Diego, CA: IEEE Computer Society.

Belsley, D. A., E. Kuh, and R. E. Welsch. 1980. *Regression Diagnostics: Identifying Influential Data and Sources of Collinearity*. New York, NY: John Wiley & Sons, Inc.

Belson, W. A. 1956. "A Technique for Studying the Effects of a Television Broadcast." *Journal of the Royal Statistical Society. Series C (Applied Statistics)* 5, no. 3:195-202

Belson, W. A. 1959. "Matching and Prediction on the Principle of Biological Classification" *Journal of the Royal Statistical Society. Series C (Applied Statistics)* Vol. 8, No. 2, 65-75. Available at: http://www.jstor.org/stable/2985543

Biggs, D., B. de Ville, and E. Suen. 1991. "A Method of Choosing Multiway Partitions for Classification and Decision Trees." *Journal of Applied Statistics* 18, no. 1:49-62.

Blyth, C. R. 1972. "On Simpson's Paradox and the Sure-Thing Principle." *Journal of the American Statistical Association* 67, no. 338:364-366.

Boulesteix, Anne-Laure. 2006. "Maximally Selected Chi-Square Statistics and Binary Splits of Nominal Variables." *Biometrical Journal* 48. no. 5:838-848.

Boulesteix, Anne-Laure. 2009. "Package exactmaxsel." http://cran.uvigo.es/web/packages/exactmaxsel/exactmaxsel.pdf.

Breiman, L. 1996. "Bagging Predictors." *Machine Learning* 24, no. 2:123-140.

Breiman, Leo. 1998. "Arcing Classifiers." *The Annals of Statistics*, 26, no. 3:801-849.

Breiman, L. 1999. "Random Forests." Statistics Department, University of California Berkeley, CA. Available at: http://leg.ufpr.br/lib/exe/fetch.php/wiki:internas:biblioteca:randomforests.pdf.

Breiman, L. 2001. "Random Forests." *Machine Learning* 45, no. 1:5-32.

Breiman, L. 2002. "Machine Learning." Paper presented at the sixty-fifth annual meeting of the Institute of Mathematical Statistics, Banff, Alberta, Canada. Available at http://www.stat.berkeley.edu/users/breiman/.

Breiman, L., et al. 1984. *Classification and Regression Trees.* Belmont, CA: Wadsworth.

Clark, P., and Niblett, T. 1987. "Induction in Noisy Domains." In I. Bratko and N Lavrac (Eds.), *Progress in Machine Learning*. Wilmslow: Sigma Press.

Cohen, William W. 1995. "Fast Effective Rule Induction." In *Proceedings of the Twelfth International Conference on Machine Learning*, 115-123.

De Comité, Francesco, Rémi Gilleron and Marc Tommasi. 2003. "Learning Multi-label Alternating Decision Trees from Texts and Data.", In *Proceedings of the Third International Conference on Machine Learning and Data Mining in Pattern Recognition*, 35-49.

De Ville, B. 1990. "Applying statistical knowledge to database analysis and knowledge base construction." In *Proceedings of the Sixth IEEE Conference on Artificial Intelligence Applications*. Seattle, Washington: IEEE Computer Society. 30-36

De Ville, B. 2007. "Ubiquitous Scoring of 1000+ Warranty Categories Using Predictive Rules Derived from Text." *Proceedings of the SAS Global Forum 2007 Conference*, Cary, N.C.: SAS Institute Inc. Available at: http://www2.sas.com/proceedings/forum2007/084-2007.pdf.

De Ville, Barry. 2008. "New Strategies for Identifying Customer Use Patterns". Proceedings of the *SAS Global Forum 2008 Conference*. Cary, NC: SAS Institute Inc. Available at http://www2.sas.com/proceedings/forum2008/118-2008.pdf.

De Ville, Barry. 2009. "Computer-implemented systems and methods for bottom-up induction of decision trees". US Patent 7,558,803, United States Patent and Trademark Office, Alexandria, Virginia. Available at http://patft.uspto.gov/.

Dobra, A. and J. Gehrke 2001. "Bias correction in classification tree construction". *Proceedings of the Seventeenth International Conference on Machine Learning*, pp. 90–97.

Doyle, P. 1973. "The Use of Automatic Interaction Detector and Similar Search Procedures." *Operational Research Quarterly* 24(3):465-467.

Doyle, P., and I. Fenwick. 1975. "The Pitfalls of AID Analysis." *Journal of Marketing Research* 12(4):408-413.

Efron, Bradley. 1979. "Bootstrap Methods: Another Look at the Jackknife." *The Annals of Statistics*. 7(1): 1–26.

Einhorn, H. J. 1972. "Alchemy in the Behavioral Sciences." The *Public Opinion Quarterly* 36(3):367-378.

Fisher, R. A. 1936. "The Use of Multiple Measurements in Taxonomic Problems." *Annals of Eugenics*. No. 7(2): 179-188.

Freund, Y. and R. E. Schapire. 1996. "Experiments with a New Boosting Algorithm." *Proceedings of the Thirteenth International Conference on Machine Learning*. 148-156.

Freund, Y. and R. E. Schapire. 2000. "Discussion of Additive Logistic Regression: A Statistical View of Boosting." The *Annals of Statistics*. 28:391-393.

Friedman, J. H. 2001. "Greedy Function Approximation: A Gradient Boosting Machine." *The Annals of Statistics* 29, no. 5:1189-1232. Available at: http://stat.stanford.edu/~jhf/ftp/trebst.ps.

Friedman, J. H. 1999. "Stochastic Gradient Boosting." Available at: http://www-stat.stanford.edu/~jhf/ftp/stobst.ps.

Friedman, J. H., T. Hastie, and R. Tibshirani. 2000a. "Additive Logistic Regression: A Statistical View of Boosting." The *Annals of Statistics*. 28:337-374.

Friedman, J. H., T. Hastie, and R. Tibshirani. 2000b. "Rejoinder for Additive Logistic Regression: A Statistical View of Boosting." The *Annals of Statistics*. 28:400-407.

Green, D. M. and J. A. Swets. 1966. *Signal Detection Theory and Psychophysics*. New York: NY. John Wiley & Sons, Inc.

Hastie, T., R. Tibshirani, and J. Friedman. 2001. *The Elements of Statistical Learning: Data Mining, Inference, and Prediction*. New York, NY: Springer.

Hawkins, D. M., and G. V. Kass. 1982. "Automatic Interaction Detection." In *Topics in Applied Multivariate Analysis*, ed. D. M. Hawkins. Cambridge: Cambridge University Press.

Ho, T.K. 1995. "Random Decision Forests." *Proceedings of the Third International Conference on Document Analysis and Recognition*. Volume 1, 278-282.

Hothorn, T., K. Hornik, and A. Zeileis. 2006. "Unbiased Recursive Partitioning: A Conditional Inference Framework." Journal of Computational and Graphical Statistics, 15(3):pp. 651-674.

Hunt, E., J. Marin, and P. Stone. 1966. *Experiments in Induction*. New York, NY: Academic Press.

Kahneman, D., P. Slovic, and A. Tversky, eds. 1982. *Judgment under Uncertainty: Heuristics and Biases*. Cambridge: Cambridge University Press.

Kass, G. V. 1975. "Significance Testing in Automatic Interaction Detection (A.I.D.)." *Journal of the Royal Statistical Society. Series C (Applied Statistics)* 24, no. 2:178-189.

Kass, G. V. 1980. "An Exploratory Technique for Investigating Large Quantities of Categorical Data." *Applied Statistics* 29, no. 2:119-127.

Kawasaki, S, et al. 2002. "Mining from Medical Data: Case-Studies in Meningitis and Stomach Cancer Domains." *Sixth International Conference on Knowledge-based Intelligent Information Engineering Systems & Allied Technologies*. 547-551.

Kononenko, I. 1995. "On Biases in Estimating Multi-Valued Attributes." Proceedings of the Fourteenth International Joint Conference on Artificial Intelligence, IJCAI 95, Montréal, Québec, Canada, August 20-25 1995, Volume 2, 1034-1040.

Lazarsfeld, P. F., and M. Rosenberg eds. 1955. *The Language of Social Research: A Reader in the Methodology of Social Research.* Glencoe, IL: The Free Press.

Lima, Manuel. 2011. "Tree of Porphyry." In *Visual Complexity: Mapping Patterns of Information.* New York: Princeton Architectural Press, p. 27.

Loh, W. Y., and N. Vanichsetakul. 1988. "Tree-Structured Classification via Generalized Discriminant Analysis." *Journal of the American Statistical Association* 83:715-725.

Loh, W. Y., and Y. S. Shih. 1997. "Split Selection Methods for Classification Trees." *Statistica Sinica* 7:815-840.

McKenzie, D. P., et al. 1993. "Constructing a Minimal Diagnostic Decision Tree." *Methods of Information in Medicine* 32, no. 2:161-166.

Meyers, Y. 1990. "Ondelettes et opérateurs. I, Actualités Mathématiques [Current Mathematical Topics]." Hermann. Paris

Michalski, R.S., I. Mozetic, J. Hong, and N. Lavrac. 1986. "The Multi-Purpose Incremental Learning System AQ15 and Its Testing Application to Three Medical Domains." In: *Proceedings of the Fifth National Conference on Artificial Intelligence*, Philadelphia, Pennsylvania, August 11-15. Volume 2: Engineering, 1041-1047.

Michie, D. 1991. "Methodologies from Machine Learning in Data Analysis and Software." *The Computer Journal* 34, no. 6:559-565.

Michie, D., and Claude Sammut. 1991. "Controlling a Black-Box Simulation of a Spacecraft." *AI Magazine* 12, no. 1:56-63.

Michie, D., D. J. Spiegelhalter, and C. C. Taylor, eds. 1994. *Machine Learning, Neural and Statistical Classification*, New York, NY: Ellis Horwood.

Miller, George A. 1956. "The Magical Number Seven, Plus or Minus Two: Some Limits on Our Capacity for Processing Information." *Psychological Review*, 63, no. 2:81-97.

Morgan, J. N., and J. A. Sonquist. 1963. "Problems in the Analysis of Survey Data, and a Proposal." *Journal of the American Statistical Association*, 58:415-434.

Neville, P. 1998. "Growing Trees for Stratified Modeling." *Computing Science and Statistics, Proceedings of the 30th Symposium on the Interface,* 30:528-533.

Pagallo, G., and D. Haussler. 1990. "Boolean Feature Discovery in Empirical Learning." *Machine Learning*, 5. no. 1: 71-99.

Pyle, Dorian. 1999. *Data Preparation for Data Mining*. San Francisco, CA: Morgan Kaufmann Publishers Inc.

Quinlan, J. R. 1979. "Discovering Rules by Induction from Large Collections of Examples." In *Expert Systems in the Micro-Electronic Age*, ed. D. Michie, 168-201. Edinburgh: Edinburgh University Press.

Quinlan, J. R. 1987. "Simplifying Decision Trees." *International Journal of Man-Machine Studies*. 27, no. 3: 221-234.

Quinlan, J. R. 1993. *C4.5: Programs for Machine Learning*. San Mateo, CA: Morgan Kaufmann Publishers Inc.

Reddy, R. K. T., and G. F. Bonham-Carter. 1991. "A Decision-Tree Approach to Mineral Potential Mapping in Snow Lake Area, Manitoba." *Canadian Journal of Remote Sensing* 17, no. 2:191-200.

Sall, John. 2002. "Monte Carlo Calibration of Distributions of Partition Statistics." SAS Institute Inc, November 18, 2002.

Schapire, Robert E. 1990. "The Strength of Weak Learnability." Machine Learning, 5(2): 197-227.

Shannon, C. E., and W. Weaver. 1949. *The Mathematical Theory of Communication.* Urbana, IL: University of Illinois Press. Republished in paperback 1963.

Shapiro, A. D. 1987. *Structured Induction in Expert Systems*. Wokingham, UK: Addison-Wesley.

Sonquist, J. A., E. Baker, and J. Morgan. 1971. *Searching for Structure (ALIAS-AID-III): An Approach to Analysis of Substantial Bodies of Micro-Data and Documentation for a Computer Program*. Ann Arbor, Michigan: Survey Research Center, Institute for Social Research, The University of Michigan.

Stone, P. J. et al. 1966. *The General Inquirer: A Computer Approach to Content Analysis*. Cambridge, Massachusetts: The MIT Press.

Surowiecki, James. 2004. *The Wisdom of Crowds: Why the Many Are Smarter Than the Few and How Collective Wisdom Shapes Business, Economies, Societies and Nations*. New York, NY: Doubleday.

Weisberg, Herbert F., Jon A. Krosnick, and Bruce D. Bowen. 1989. *An Introduction to Survey Research and Data Analysis*. 2d ed. Glenview, IL: Scott Foresman.

Weiss, S. M., and C. A. Kulikowski. 1991. *Computer Systems That Learn: Classification and Prediction Methods from Statistics, Neural Nets, Machine Learning, and Expert Systems*. San Mateo, CA: Morgan Kaufmann Publishers Inc.

Wikipedia. "Artificial Intelligence." Citation of McCarthy John. 1955. http://en.wikipedia.org/wiki/Artificial_intelligence#cite_note-Coining_of_the_term_AI-2.

Index

C

R

S

Z

Symbols and Numerals

CPSIA information can be obtained at www.ICGtesting.com
Printed in the USA
LVOW03s1133241014

410287LV00003B/35/P